THE BACKYARD HOMESTEAD SEASONAL PLANNER

THE BACKYARD HOMESTEAD SEASONAL PLANNER

WHAT TO DO & WHEN TO DO IT

Ann Larkin Hansen

Storey Publishing

The mission of Storey Publishing is to serve our customers by publishing practical information that encourages personal independence in harmony with the environment.

Edited by Deborah Burns and Sarah Guare
Art direction and book design by Michaela Jebb
Text production by Erin Dawson
Indexed by Christine R. Lindemer, Boston Road Communications

Cover illustration by © Michael Austin
Interior illustrations by © Elara Tanguy, except pages 29 (bottom 3), 68, 77, 125, and 152 by Bethany Caskey and page 110 by Elayne Sears
Maps pages 4 and 5 by Ilona Sherratt

Storey Publishing
210 MASS MoCA Way
North Adams, MA 01247
storey.com

Printed in China by R.R. Donnelley
10 9 8 7 6 5 4

Library of Congress Cataloging-in-Publication Data

Names: Hansen, Ann Larkin, author.
Title: The backyard homestead seasonal planner : what to do & when to do it in the garden, orchard, barn, pasture & equipment shed / Ann Larkin Hansen.
Description: North Adams, MA : Storey Publishing, 2017. | Includes bibliographical references and index.
Identifiers: LCCN 2017011752 (print) | LCCN 2017013970 (ebook) | ISBN 9781612126982 (ebook) | ISBN 9781612126975 (paper w/partially concealed wire-o : alk. paper)
Subjects: LCSH: Farms, Small. | Seasons.
Classification: LCC HD1476.A3 (ebook) | LCC HD1476.A3 H35 2017 (print) | DDC 630—dc23
LC record available at https://lccn.loc.gov/2017011752

Contents

Preface

When my family and I first moved to this farm, I wanted to do it all: garden, keep an orchard and bees, grow field crops, raise livestock and poultry, have my own firewood and timber. It was a steep learning curve, though my job as a reporter for a weekly farm newspaper helped a lot, since it took me to several hundred farms and introduced me to many farmers and agricultural professionals. There were plenty of mistakes and some outright failures along the way, but I did learn, and my husband and I are still here and still happy, rejoicing in homegrown food and even making a little money most years.

The three biggest things I've learned over the past 25 years are:

YOU CAN'T DO IT ALL AT THE SAME TIME if you have a small budget, a family, and outside job obligations — not if you want a decent life and good family relations.

YOU DON'T NEED TO DO IT ALL. What does work nicely is to try different enterprises — pigs, bees, timber harvests, whatever — over the years to discover what you like best, whether they fit together well, and what most benefits your family, your land, and your budget. Keep what works.

DOING A THING IN ITS PROPER SEASON — when it would naturally occur, or when conditions make the job most efficient and comfortable — is how you spread the work more evenly through the year. This is what makes it possible to get it all done and still have some leisure time.

This book is about timing, as important an aspect of raising plants and animals as any other. Bits and pieces of information on when to do farm-related tasks are widely available in gardening calendars, vaccination schedules for livestock, discussions of when to prune trees and plant crops, and so on. This book combines all of this information to create a seasonal calendar for the most common enterprises found on small diversified farms, in a way that spreads the work appropriately and more or less evenly and can be understood and applied anywhere in the temperate zones of this continent. This book assumes that you wish to work with natural processes instead of against them, and so the calendar of seasonal tasks in each chapter excludes the use of synthetic chemicals.

Every rural acreage is different in what and how and when to do things, because every piece of land is different and so is every landowner. For that reason, no one else (and especially no book) can possibly tell you precisely what and how and when *you* should do things on *your* place. A book tells you what to look for and which questions to ask. You have to figure out the answers.

This book is just a beginning; it gets you to the starting line. From there you have to run the course: do the thinking and planning and experimenting over the years to fill out the details of your own personal farm calendar. The people who do that — think and plan and experiment — are the ones who thrive.

Introduction

Defining the Seasons

Seasons vary a lot, year to year. They're wet or dry, cold or warm or hot, early or late. Just now we're in the middle of the most drawn-out mud season, otherwise known as late winter, that I've ever seen, and it's been a challenge to keep the cattle even reasonably dry.

The timing and length of the seasons differ strikingly across temperate North America, but the seasons themselves are similar enough that we recognize the same broad characteristics. Winter is a time of cold and no plant growth; summer means heat and long days. Spring is the transition from winter to summer, and fall the transition from summer to winter. Each season slides into the next; only rarely is there a sudden shift. In the most southern states, winter almost disappears and summers are hot enough to keep cool-loving plants from growing; in the northern tier of states, summers are short and relatively cool and the growing season isn't long enough for tomatoes and other long-season plants to mature when started outside.

As a culture, we've gotten into the habit of looking at the calendar to determine planting times and other such decisions, no matter what's going on outside. But a few generations ago, instead of looking at the calendar, farmers and gardeners looked at what the local plants and birds and other wildlife were doing — a much more accurate indicator considering how different seasons are from year to year. Using plant growth and wildlife activity to identify how far along you are in the year is called phenology, and it is a highly useful study for anyone who grows crops or raises animals.

But phenology is a very local study, since seasonal indicator species vary widely across various regions. Spring in the Southwest has a different cast of native characters than spring in the Northeast. The signs can vary even across small areas. Our friends just 15 miles to the south know that sprouting skunk cabbage means winter is coming to an end, but I've never seen it in our neighborhood. I watch for returning red-winged blackbirds instead. Phenology cannot define what makes a season across the entire temperate zone of the continent.

Useful Seasonal Indicators

There are, however, some precise and fairly universal indicators of the different seasons that apply no matter where you live.

SOIL AND AIR TEMPERATURES. These can be measured with air and soil thermometers (soil temperature is measured at a depth of 3 to 4 inches) or by keeping an eye on what local farmers are doing. There are three crops grown throughout North America — corn, wheat, and alfalfa — that, no matter where they are planted, germinate and grow best with specific soil and air temperatures. Wherever you are in the country:

- Spring wheat goes in the ground in early spring.
- Most farmers plant field corn in the middle of spring.
- The first cutting of alfalfa for hay occurs in late spring into early summer.

Soil temperature is measured with a soil thermometer at a depth of 3 to 4 inches.

Warm- and Cool-Season Grasses

Cool-season grasses commonly used for forage: Tall fescue, Kentucky bluegrass, orchard grass, redtop, smooth bromegrass, timothy

Warm-season grasses commonly used for forage: Big bluestem, Bermuda grass, Caucasian bluestem, eastern gamagrass, Indian grass, little bluestem, side-oats grama, switchgrass

Note: Information adapted from USDA Natural Resources Conservation Service Grazing Fact Sheets.

RATE OF GROWTH IN WARM- AND COOL-SEASON GRASSES. This is a second good and widespread indicator of the season. Except for the first cutting of alfalfa, harvest times are not reliable indicators — crops get harvested when farmers judge that they're mature, more so than according to the season.

In this book, I have used soil and air temperatures and grass growth as indicators to generally define 12 seasons: an early, middle, and late period of each of the 4 major seasons. This level of detail allows for a much more specific discussion of what should best be done at a particular time of year, and why.

In Season and Out

If you have enough money, labor, equipment, and infrastructure, it's certainly possible to do many things out of season. We can grow tomatoes in North Dakota in winter with a warm and well-lit and expensive greenhouse, calve beef cattle not just in spring but in August or January, and cut firewood during the worst heat of summer. But why? For the small-scale diversified rural acreage, this is neither practical nor enjoyable. I don't want to work that hard. Doing things at the right season requires less money and labor, and it produces better results.

On the other hand, it's a rare year when I get even most of my work done at exactly the right time. Sometimes the weather just won't cooperate; other times some other project demands all my attention, or there's a major family event involving travel that I don't want to miss. Sometimes things have to get done at less than the best times, and sometimes they wait until next year, like getting those grapevines off the field fences. I've realized that if I shoot for the best and am willing to settle for what works, we'll get by okay.

Seasons Defined

This book generally defines the seasons as follows:

Midwinter. Soil is frozen or subject to intermittent freezing. Air temperatures are continually or often below freezing. No plant growth. In USDA climate Zones 1–7, the soil is on average frozen in winter; in Zone 8 it's subject to freezing. In Zones 9 and 10, the soil does not normally go below 40°F/4°C, and air temperatures only rarely go below 32°F/0°C, so that these areas can be said to have no real winter.

Late winter. Soil surface thaws in the sunniest areas, but its average temperature is still below 40°F/4°C. Daytime air temperatures are frequently above freezing. No grass growth, though the very earliest native plants begin to emerge.

Early spring. Soil thaws and warms through the season to 40°F/4°C and higher. Cool-season grasses begin to grow, but slowly. Spring small grains are planted as soon as the ground can be worked.

Mid-spring. Soil temperatures reach 50°F/10°C and continue to warm through the season. Cool-season grass growth accelerates; field corn is planted.

Late spring. Soil temperatures reach 60°F/16°C and warm through the season. Cool-season grass growth explodes; soybeans, sunflowers, and sorghum are planted. First cutting of alfalfa gets under way.

Early summer. Begins after the average date of the last hard frost. Soil temperatures warm to 70°F/21°C or higher. First cutting of alfalfa is finished.

Midsummer. Cool-season grass growth slows (when daytime air temperatures are over 75°F/24°C) or stops completely (when temperatures are over 90°F/32°C). Warm-season grasses grow rapidly (best at 80 to 90°F/27 to 32°C).

Late summer. Daily high air temperatures begin to cool off and cool-season grasses begin to grow again. Warm-season grass growth slows.

Early fall. Soil temperatures cool down to 70°F/21°C or below. Cool-season grasses show a burst of growth. Warm-season grasses stop growing.

Mid-fall. Begins with the first hard frost, which kills the corn. Deciduous trees shed leaves; average daily air and soil temperatures trend to below 60°F/16°C; cool-season grass growth slows and stops.

Late fall. Daytime soil temperature falls below 50°F/10°C; nighttime freezing air temperatures are frequent.

Early winter. Soil temperature falls below 40°F/4°C. Plant growth ceases.

Average Timing and Depths of Frost in the U.S. and Canada

When you're mapping out a work schedule and organizing your equipment and supplies for the seasons ahead, it's helpful to know the average dates of first and last frost in your area. But bear in mind that the dates in the calendars below are only averages, and the actual dates can vary by several weeks from year to year. The average dates for field work in your state can be found online in the USDA's National Agricultural Statistics Service Handbook #628, "Usual Planting and Harvesting Dates for U.S. Field Crops." Especially useful for planning purposes are the dates for planting spring grains and field corn, respectively, and the first alfalfa harvest. Again, bear in mind that these dates are only averages and will also vary by a couple of weeks or more in states that have big changes in elevation (Arizona, for example), or that are very long from north to south (such as California and Minnesota).

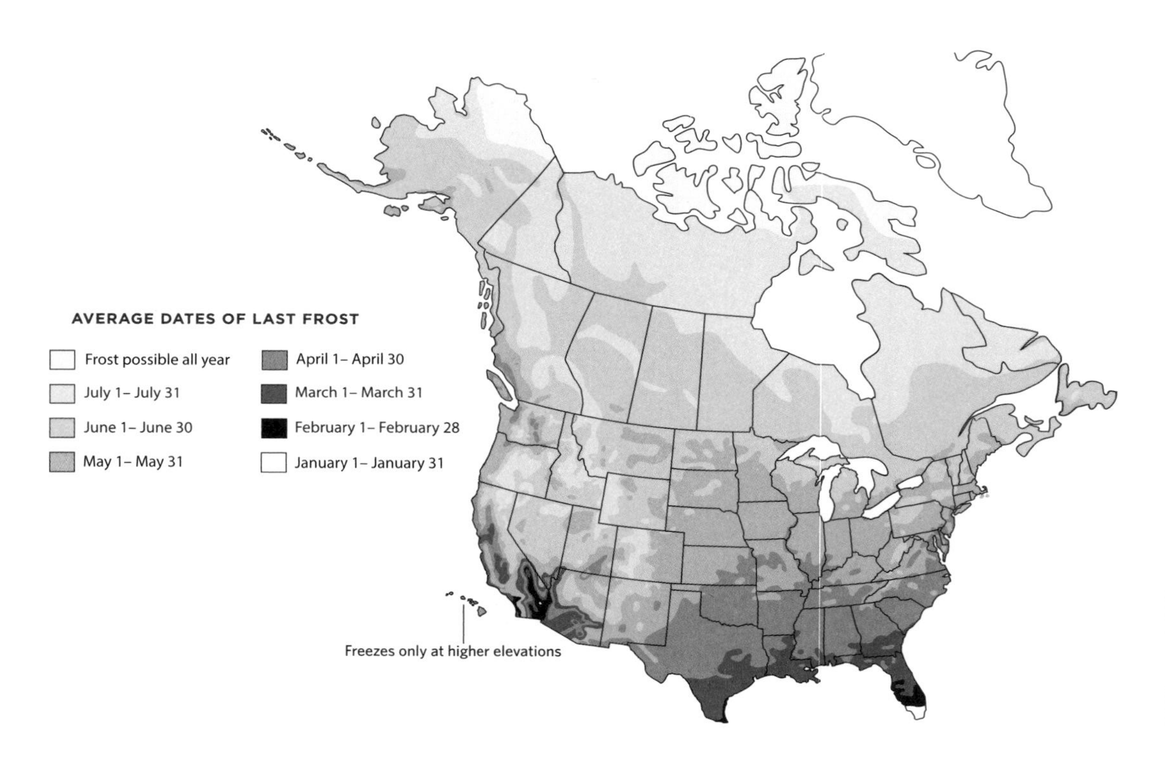

AVERAGE DATES OF FIRST FROST
Frost possible all year
August 1–August 31
September 1–September 30
October 1–October 31
November 1–November 30
December 1–December 31

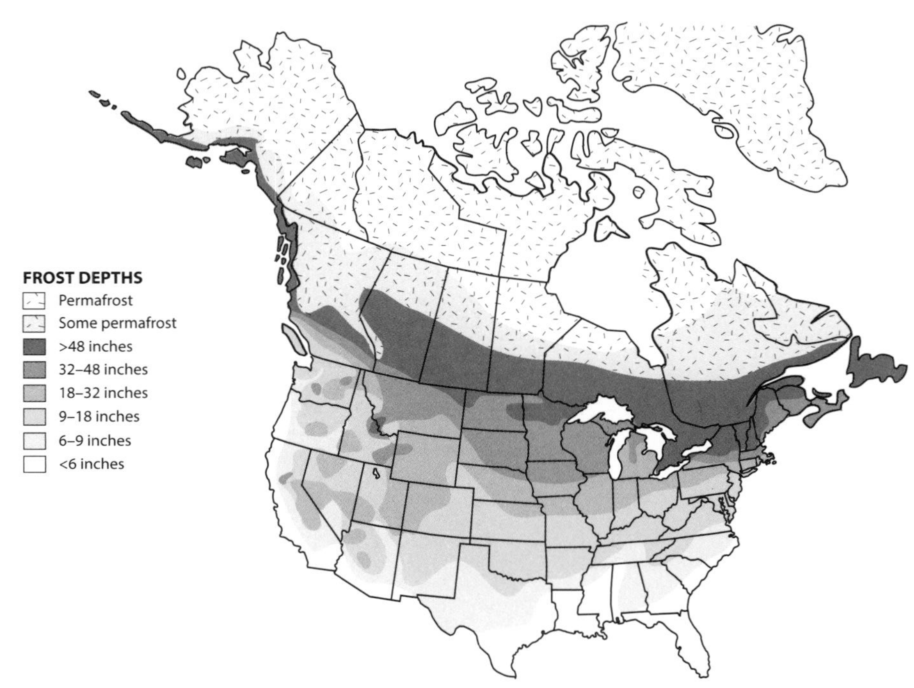
FROST DEPTHS
Permafrost
Some permafrost
>48 inches
32–48 inches
18–32 inches
9–18 inches
6–9 inches
<6 inches

CHAPTER ONE

Midwinter

Soil is frozen or subject to intermittent freezing. Air temperatures are continually or often below freezing. There is no plant growth. In USDA climate Zones 1–7, the soil is on average frozen in winter; in Zone 8 it's subject to freezing. In Zones 9 and 10, the soil does not normally go below 40°F/4°C, and air temperatures only rarely go below 32°F/0°C, so that these areas can be said to have no real winter.

MIDWINTER IS THE SEASON of short days and long nights between the winter solstice and when the first warming trend starts sap moving in the trees — the coldest time of year. The soil is cold or frozen, and nothing grows. Aside from livestock chores and filling the woodbox, there isn't much that needs to get done around the farmyard; it's a traditional time to make and mend tools and pursue other indoor crafts. This is also the traditional time to pour a cup of coffee and do the annual recordkeeping, taxes, and ordering of seeds, supplies, chicks, bees, and equipment parts.

Midwinter is conference season, too, since it's when farmers have some spare time on their hands. Watch for multi-topic conferences on gardening, homesteading, and sustainable farming, as well as workshops on everything from sheep shearing to fruit tree pruning and beekeeping to rotational grazing and direct marketing. Besides the information you gain, these sessions can be a great way to meet other people interested in the same things you are.

The one big outdoor project of the season is woodlot harvests and improvements. This is traditionally the height of the logging season in regions where the ground freezes, though in areas where the ground doesn't freeze, no heavy woods work is done unless the ground is *dry*, in order to minimize soil compaction and erosion and maximize safety for the workers. Woodlot owners can always find plenty to do: thinning and pruning trees, piling or burning debris, making firewood, and harvesting timber for various purposes.

Midwinter

FOCUS ON:
paperwork • woodlot

Garden

» **Read your notes** from the past growing season to assess how well the previous year's varieties, rotation, and cover crops performed.

» **Use your notes** to plan this year's crop rotation and cover crop plantings, then order seed.

» **Price and order** or purchase soil amendments (based on soil test results) if not applied last fall, for spring application.

Field

» **Assess how well** the previous year's crops performed by going over your notes from the past growing season.

» **Plan this year's crop rotation** and cover crop plantings based on your notes from last year, then order seed.

» **Purchase soil amendments** (based on soil test results) if not applied last fall, for spring application.

Pasture

» **Order seed** for overseeding winter feeding areas in early spring.

» **Work on clearing fence lines** of brush and branches if there's little or no snow.

Orchard

» **Select and order new trees** and pest controls — traps and treatments — if you want them.

» **Sprinkle wood ashes** around trees at the dripline.

Beeyard

» **Observe hives** on warm, sunny days (over 45°F/7°C), looking for bees emerging for "cleansing flights." Do not open hives.

» **Order hives,** supplies, and bees for spring delivery.

» **Assemble and paint hives** when they arrive.

Barn

» **Check the hay supply:** you should have at least half left. If you're short, start looking for more to buy.

» **Check all livestock,** and monitor water, salt, mineral, and hay feeders daily.

» **Ultrasound goats and sheep** to see how many young they are carrying, if desired.

» **Control rodents.**

» **Research where you can buy** weaner pigs in spring.

Coop

- **Order chicks** for delivery in early to mid-spring.
- **Check feed,** and water hens daily.
- **Let hens outside** on nice days.

Equipment Shed

- **Clean and lubricate** any tools and equipment that are in use.

Woodlot

- **Harvest timber** and do forest improvement work (pruning, thinning, debris cleanup) in regions where the ground is frozen or dry.
- **Burn debris** if there is snow on the ground or if the weather is wet.
- **Check over your sugaring equipment** if you have sugar maples and plan to tap them.

Wildlife Habitat

- **Research what native plants** and animals occupied your acres before settlement.
- **Build birdhouses and bat and wild bee houses.**
- **Order native plant seed,** food trees, and shrubs for wildlife.

Garden

By this time of year, you should have a nice pile of seed and plant catalogs or have identified some promising seed companies on the Internet. Good catalogs will describe each variety's preferences and tolerances for heat, cold, and drought as well as number of growing days to harvest, growth habit and mature height, disease resistance, and other important details. When deciding which species and varieties to plant, start with the following checklist:

- **Choose varieties suited to your climate and length of growing season.** For example, summers in the Southeast are blisteringly hot and humid, in the coastal Northwest it's cool and moist, and where I am, in the upper Midwest, the average summer is warm and humid, with a relatively short growing season.

- **Look for varieties that are disease resistant** if you have persistent disease problems (fusarium wilt is a common one).

- **Note how the seed is produced if you have opinions about this.** Note whether it is open-pollinated, organic, hybrid, or genetically modified (GM). This is especially important if you plan to save your own seed (for more information, see Deciphering Seed Labels, page 11).

- **Match perennials with your hardiness zone or provide a microclimate.** If a perennial, tree, or bush is rated for a zone warmer than yours, it's probably not going to thrive or survive, though you may have success by utilizing a microclimate: try planting it against a wind-protected south-facing wall and mulching the root zone after the ground freezes in early winter.

If you planted last year, use your notes to plan the details of this year's crop rotation and cover crop plantings. Assess how well the previous year's varieties, rotation, and cover crops performed in terms of:

- Quality and productivity
- Resistance to weed competition, disease, and insect pests
- Ease of harvesting
- Contributions from the roots and aboveground residue to soil protection and health

Deciphering Seed Labels

Open-pollinated. Seed from plants that handled their own breeding naturally in the field. These strains have been stabilized over the years so they "breed true." Seeds that are saved from these strains will produce plants like the parent.

Heirloom. An open-pollinated plant (or domestic livestock breed) that has been bred for specific traits for more than 50 years, especially hardiness and flavor, and is often specifically adapted to a particular region.

Hybrid. The result of crossing two varieties of a single species of plant (or occasionally between two species) in the field. Seed from hybrid plants does not breed true and may be sterile.

Genetically modified organism (GMO). A plant created *in the laboratory* by inserting genes from one plant into another *in a way that would not occur in nature*. This technology allows breeders to cross-plant species with unrelated species and even nonplant species. Genetically modified (GM) seed is not generally sold to home gardeners, but it probably will be soon. GM seed is common in commodity crops. GMOs are generally protected by patent, and the seed cannot legally be saved. GMOs are prohibited in organic production. Companies do not have to disclose in their seed descriptions whether their seed is genetically modified.

Cell fusion CMS (cytoplasmic male sterility). A particular method for genetically altering seed, creating a type of GMO. The modified trait is not passed on to the seed's progeny and is therefore allowed in National Organic Program–certified organic seed production.

Treated seed. Seed coated with fungicide and/or pesticide to keep it from rotting if planted when soil moisture or temperature inhibits rapid sprouting. Prohibited in organic production.

CRISPR (clustered regularly interspaced short palindromic repeats). A type of GMO. A cheap, effective, and (mostly) precise method for editing DNA in the laboratory, and therefore a type of genetic modification. The gene modification is permanent and passed on to all future generations. CRISPR seed is expected to be commercially available in the near future, if it isn't already by the time this book is published.

Note: This information is courtesy of Jamie Chevalier, catalog editor, *Bountiful Gardens*.

What's the Difference between a Garden and a Field?

Though there's no hard-and-fast separation between the two, I think of a garden as supporting a variety of crops for consumption in a relatively small area, while a field grows a single crop at a time (though this can be combined with green manures and cover crops) in a larger area.

Traditionally we tend to think of gardens as producing vegetables, herbs, and (nontree) fruits, and fields as producing grains and forages. But throughout the United States, there is field-scale vegetable growing as well as plenty of grain patches in gardens. Fields, we would posit, are large enough to make mechanized implements essential for efficient production, while a small garden can be managed with hand tools.

Field

Though field crop production is not common on small diversified acreages, it's helpful to understand what your farming neighbors are doing through the year so that you pick up on seasonal indicators. If you do have field crops, now is the time to plan your rotations and order seed.

Field crops are either row crops or nonrow crops. Row crops are grown with enough soil between the rows so plants aren't crowded and you can get machinery through for cultivating (mechanical weeding) and harvesting. Corn, soybeans, cotton, and vegetables, for example, are all grown as row crops. Nonrow crops, most typically small grains (wheat, oats, rye, barley, and the like) and forages (grasses and legumes for hay), are planted as a solid stand, so you can't get equipment in or even walk in the field without treading on the crop. Cultivation is not possible, so weed control depends on other methods. If you're organic, the most important thing is to achieve good weed control in the previous year's crop. (For more about weed control, see page 143.)

When ordering seed for your fields for the coming season, do an Internet search or ask neighbors to identify seed dealers in your region. Keep in mind the same concerns about varieties and origins as for your garden seeds (see Deciphering Seed Labels, page 11). Calculate the amount of seed to order for row crops based on how the plants will be spaced in the row and how much space will be between rows. There are no strict spacing rules for any crop; the goal is to have spacings appropriate for your equipment and the water, sun, and soil nutrient needs of your specific crops without leaving too much room for weeds. Nonrow crop seed amounts are based on the recommended bushels per acre for your location, your soil type(s), and the variety you're planting. If you're combining crops, such as brome grass and alfalfa for a mixed hay, you'll need to decrease the seeding rates for each type of plant.

Row crops are placed at the appropriate soil depth with a planter, while nonrow crops are either broadcast and then lightly tilled or pressed into the soil, or drilled (planted under the surface) with a grain or grass seed drill.

For organic farms, the USDA National Organic Program (NOP) Final Rule requires that at least three different crops be grown over a five-year period, with at least one of those crops being a leguminous green manure. The rotation should utilize both soil-building and erosion-preventing cover crops and mulches. This means your seed order doesn't end with the selected crop variety. Appropriate cover crop seeds will be needed, too, and a plan for when these get planted and tilled under. For example, you could overseed red clover into a stand of field corn when it's 8 inches high and leave the clover there through winter before tilling it under in early spring ahead of planting pumpkins. The possibilities are numerous, given the variety of both crops and covers that can be grown, their different effects on soil nutrient and organic matter levels, when the production crop needs to be planted and harvested, the need for winter soil cover, and to have something that's easy to till under in spring or that will winterkill so you can plant the next crop directly into the cover crop residue. For more on this subject, see page 82.

Pasture

Pastures are resting at this season. If animals are being outwintered (fed outside through winter) on pasture or hayfield and you're using electric fence around the hay stockpile, check the fence regularly and especially after storms, when ice and heavy snow may have stretched or broken wires. (For more about outwintering livestock, see page 152.)

On the paperwork side, this is a good time to review last year's rotation paddocks and schedule and to plan any changes or refinements for the coming season (for details, see page 66). If you noticed areas of poor growth or that the winter feeding area is more mud than grass, then order seed for replanting these areas in spring. Also look over your fencing supplies and order items that are unavailable (or that need to be better quality than what is available) at the local farm store — this applies especially to good electric fence wire, posts, and energizers.

Orchard

If you're starting an orchard, this is the time to research and order trees that are suited to your climate and your personal tastes (see page 112 for details).

Spread wood ashes around the trees at the dripline to add potassium and micronutrients to the soil. If the snow is deep, you can pack it down around the trunk guards to discourage mice from tunneling in to chew the bark. This is also a good time to clean and sharpen pruners, loppers, and chains for the small chain saw, since pruning begins next season.

Beeyard

Hives should be observed but not opened. On the warmest days (above 45°F/7°C), bees should be seen removing dead bees and taking short "cleansing flights" (to eliminate wastes). These activities and a hum from the hive, where the bees are in their winter cluster, indicate health. Dead bees collected around the opening, a weak hum, and a light hive indicate problems. But since the bees sealed the cracks in the hives last fall to protect themselves against drafts, don't open hives; it will chill the bees. Provide food close by if the hive feels light (tilt the hive gently to get a feel for this). Do make plans for addressing any possible problems in spring.

Barn

When you go out each morning to feed and water, look over all livestock (as you should year-round) with an eye to their overall condition and for any signs of distress such as shivering, lameness, or coughing. Animals that are thin, have dull coats, dull eyes, signs of heavy coat licking, or a lack of interest in their feed may be suffering from internal parasites, lice, contagious disease, or an infection and should be diagnosed and treated. Thin animals may simply need a richer diet to deal with cold weather.

At the middle of midwinter, you should have half or more of your hay supply left. If it looks like you're going to be short, buy as soon as possible — supplies dwindle and prices rise as spring approaches.

Cattle, horses, and sheep can be outwintered or kept in a dry lot with a nearby shed for shelter from wind, rain, and heavy snow. Pigs and goats in general do not tolerate cold well and should be well bedded and inside in severe weather, though you may let the goats out on nice days.

Pregnant ewes and does are often ultrasounded at this season, to discover how many young they are carrying so that their feed ration can be adjusted accordingly in late pregnancy. Find a mobile ultrasound service by contacting area goat and sheep groups. Also find a source for weaner pigs in spring.

Coop

Even in very cold weather, chickens (natural heat lovers) hate being cooped up; give them access to the outdoors except in extreme weather. They may not go far (most chickens I've known detest snow).

Midwinter is the time to order new chicks, since hatcheries need a bit of lead time for eggs

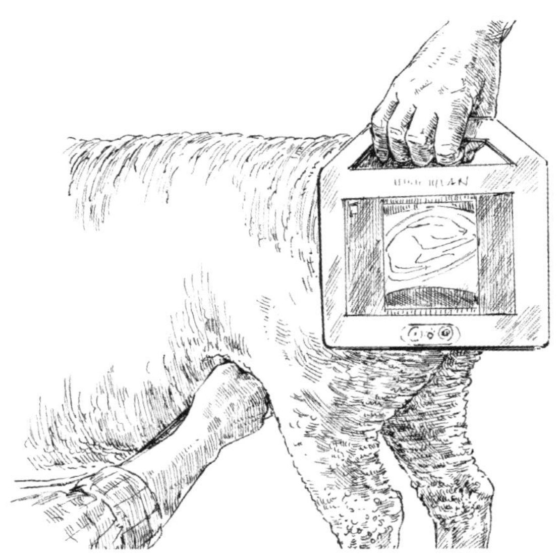

A portable ultrasound unit requires some training to use, but it is handy for assessing the stage of pregnancy and the number of lambs.

to be laid and hatched for spring delivery. The order form should allow you to choose the week you want the chicks delivered; if you can, time it so the new laying hen chicks will start to lay as your older birds are starting to molt and not laying. This will depend some on the breed you buy, since maturity rates vary. In general, you can figure about five months from chick to egg layer, and that hens will start molting in late summer. If that has your chick order arriving in late winter, you can either set up a brooder in the house until the weather warms (and put up with the extra dust and smells) or save up eggs in midsummer during peak egg season to get you through the molting gap.

Meat birds should be started early enough that they'll be feathered out and ready to go on pasture in mid-spring, when the weather is moderate and the grass is growing well, and ready to butcher in midsummer. Or start them in midsummer for butchering in early fall. You can butcher at anywhere from 8 to 12 weeks of age, depending on the breed and what you want your finished weight to be. Cornish-Rock crosses are the fastest maturing; heritage breeds take several weeks longer to reach similar weight.

Equipment Shed

If you didn't do so last fall (and I usually don't get around to it then), buy or order parts specific to each piece of equipment that often have to be replaced during the season. Belts, chains, and shear bolts are the most common.

Midwinter is a good time for the small, more creative projects that were put on hold until you were less busy: making a leather holster for the pruning shears, figuring out how to hook up your bike to the grain mill, building a better tomato cage, or learning how to sharpen your own chain saw blades.

Woodlot

The bugs are gone, there's not much going on around the homestead, and the weather is cold enough to make heavy work in heavy clothes a relatively sweat-free task. It's the perfect time to be in the woods. Even if you have only a small patch of trees or just a few in the yard, it's time to thin and prune if needed, or cut down and remove any that are diseased or starting to die (see page 156).

If you have sugar maples on your property and make syrup, this is the season to get your maple syrup equipment out of storage and checked over. Also make sure you have plenty of dry wood stacked for boiling the sap.

Mark trees to be removed with marking paint or flagging tape.

Rats, Mice, and Pigeons

Winter is a good time to control rats and other pest animals. Food supplies are limited, which causes them to stick closer together than usual — making them easier targets. Cats are excellent at controlling mice; make sure they have access to the garage, too, since mice like to nest in the engine compartment of cars and chew on the wiring. But rats are a different story.

The only effective way we've found of getting rid of rats is poison. The hardest and most important thing about using poison is keeping it away from livestock and pets. If you can't place the poison inside the walls or under the floors where the rats live, then put it out after all pets and livestock are locked in for the night and remove any leftovers before you let your animals out in the morning. Rats almost always go into their burrows to die, but nonetheless keep an eye peeled for corpses and pick them up before pets find them. Keep putting out poison until all signs of rats are gone.

Rats can appear quite suddenly — the old-timers say that if someone tears down an old shed within a couple of miles of your barn, you'll get rats. After a couple such incursions, we cemented in all the holes and cracks in the barn floor and tore down the ancient chicken coop — and we haven't had rats since. But that's not saying they won't turn up again.

Pigeons are easier to shoot than to poison if you have livestock, since it's hard to find a good spot in the open to leave any poison where your animals can't get at it as well, and there's always the danger that your dogs or cats will find a poisoned dead bird and eat it. Shoot pigeons at night, when they're roosted in the rafters of the barn and outbuildings. Use an air rifle or lightly loaded birdshot — something that won't damage the roof when you miss. Even more effective is to close off all entrances where pigeons can get into buildings, though this can be difficult in haylofts or large sheds, where big ladders are needed to get to the high spots.

Some tips: If trails become icy, put woodstove ashes on them to increase traction. Keep rolls of red and green flagging tape in your pocket to mark trees that should come down (red is dead) and those that should be given more space (green is grow). When you come back with equipment, you won't waste time trying to figure out what you want to do.

Working with the Weather in the Woods

In midwinter, woods projects can be strategized as follows:

ANYTIME is a good time to prune and thin, unless tree diseases are a problem in your area (oak wilt is a big concern in our region). If so, wait until the weather is cold enough to prevent their transmission (generally below 40°F/4°C).

IF THE GROUND ISN'T FROZEN, uproot invasive species and dispose of them by composting or drying and burning when conditions are right. Woody invasives especially can be piled to dry and burned later. In mild winters a snow may fall and insulate the ground before it starts to freeze, and you can uproot plants even through the snow.

WHEN THERE'S LITTLE OR NO SNOW, pick up pruned branches and excess debris on the ground and stack for later burning or build brush piles. Thin stands of young trees, and stack or pile the thinnings. Flag and cut unhealthy trees. With the leaves down and the ground bare or nearly so, this is an excellent time to lay out new trails since you can see the lay of the land so well.

WHEN THERE'S ENOUGH SNOW TO HIDE DEBRIS AND THE BASE OF TREE TRUNKS, you may want to quit felling trees. It can be difficult to leave short stumps, since putting your chain saw down through the snow to the base of the tree means you can't see any rocks or debris in the way that might suddenly dull the blade, or worse, cause it to kick back. If you really want to keep on cutting trees, stamp the snow down around the base of the trunk before cutting, so you can see what you're doing. Keep pruning, and burn debris.

WHEN IT'S WET IN THE WOODS, you can burn piled debris or go to the woodshed and split wood. Wet soil and leaves are slippery and dangerous to work on when chainsawing. Travel woods on foot when ground is soft, since trails rut easily from equipment traffic.

Wildlife Habitat

While you're researching and ordering seeds, supplies, plants, and livestock, do a little additional research and find out what plant and animal communities occupied your land before it was settled. Start with your state's Department of Natural Resources (or its equivalent) and local conservation organization websites. Then plot out where remnants might exist on your acres, or locate areas where you could reestablish some of these species. Order native plant seeds, shrubs, and plants. Especially check with state and federal government agencies for lower-cost seed and plants; the selection may not be as good, but the price is usually right. If you're planning a major project, such as a prairie or wetland restoration, or if you are planting a lot of trees, research what technical and cost-sharing help is available from agencies such as the Natural Resources Conservation Service, U.S. Fish and Wildlife Service, and state agencies.

Topic of the Season

Soil Health and Fertility

Soil is the heart of our land, where the health and productivity of plants, animals, and humans are all centered and connected. A soil's type, fertility, and condition (ability to hold and dispense moisture, nutrients, and air to plant roots) determine whether the species of plants and animals we want to raise will thrive on that ground. The first concern of the sustainable gardener and farmer is to improve and maintain soil health and fertility.

This is the time of year to educate yourself by reading texts about soil fertility and by reviewing your soil tests. Soil tests are the foundation of your soil-building program: use them to determine what soil amendments will be added when (and order anything that needs to be ordered), and which cover crops/green manures will best address soil needs and how they will be worked into crop rotations.

Soil Type

The primary component of soil is rock particles, which come in three basic sizes: sand (0.05 to 2 mm in diameter), silt (0.002 to 0.05 mm), and clay (less than 0.002 mm). The proportions of each determine soil type, as shown in the diagram at right. The best all-around soil for producing human and livestock food is loam, technically a mix of 30 to 50 percent sand, 25 to 50 percent silt, and 10 to 25 percent clay. The soils of the entire United States have been mapped and classified. Use the interactive service on the Natural Resources Conservation Service website (see Resources) to find your land's soil types and classifications, or visit your county agricultural Extension office to get a map of soil types and classifications for your land.

Sandy soils are light, and so are easier to work with either hand or mechanized implements. They are quick to both warm up in spring and dry out in summer, and they do not hold nutrients well. To maintain fertility in these types of soils, frequently apply smaller amounts of amendments and organic matter.

Clay soils are heavy to work, slow to warm up, and slow to dry out. Because they hold nutrients well, you can generally apply amendments in larger amounts less frequently. Clay needs lots of organic matter to keep it loose.

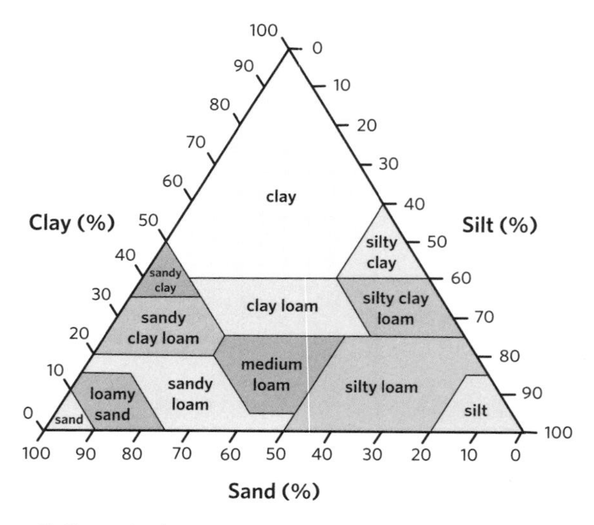

Soil type is determined by the relative proportions of differently sized rock particles.

Soil Class

Soils throughout this country are classed by their suitability for agriculture:

- **Class I and II** soils have few limitations.
- **Class III and IV** soils have severe limitations that may reduce what you can grow or require careful conservation practices.
- **Class V, VI, and VII** soils are not suitable for cultivation and so are usable only for grazing, forestry, and wildlife.
- **Class VIII** soils are not suitable for any commercial purposes.

In addition, soils in every class can be sorted into several subclasses:

- **Subclass "e"** soils are prone to erosion.
- **Subclass "w"** soils have poor drainage and/or are subject to flooding.
- **Subclass "s"** soils have rooting zone limitations.
- **Subclass "c"** soils are limited by climate.

Windbreaks, contour or terrace cropping, grass strips, and earth berms are used to limit wind and water erosion; surface or subsurface drainage may fix some waterlogged soils. But you can't easily stop flooding or repair severely alkaline soils.

Soil Fertility

For good growth, food plants for livestock and humans need to absorb 13 nutrients from the soil, in somewhat differing proportions depending on the species. Nitrogen, potassium, and phosphorus are generally needed in the largest amounts, followed by calcium, magnesium, and sulfur, with relatively minor (but still very important) amounts of boron, chlorine, copper, iron, manganese, molybdenum, and zinc. Some of these nutrients are present naturally in soils due to the parent

Organic Soil Amendments

A regular diet of compost, animal manures, and green manures is all most soils will need to stay alive and fertile. If a soil test indicates a pH that's too high or too low, add lime or sulfur at the recommended rate. If the test indicates some nutrient deficiencies too severe to be corrected by compost and manure, the job is to find a natural amendment that will fix the problem. You can search online or consult a sustainable and/or organic farming or gardening association in your area. In general, bone-, blood, feather, fish, and hoof meal are generally good slow-release sources of nitrogen; wood ash is excellent for potassium (don't put it on too heavily); and for phosphorus, use bonemeal. Dolomitic lime adds both calcium and magnesium and raises pH; if the pH level is okay or a little high, use gypsum instead. For trace minerals, dried seaweed is a good place to start.

rock; for example, soils derived from limestone are high in calcium. To supply nutrients that are insufficient or absent according to a soil test, apply purchased amendments, green or animal manures, or compost. (For a discussion of cover crops and green manures, see page 82.)

Soil Condition

Soil condition is the soil's ability to hold nutrients, as well as both water and air, in amounts beneficial to plants. Soil in good condition ("in good heart") will crumble into clumps in your hand when it's moist, forming neither a single clump nor a pile of dust. Soil condition depends in large part on the percentage of organic matter (anything that was once living and now isn't), which is ideally between 2 and 6 percent. Also of critical importance is the amount and diversity of the living microorganisms that break down organic matter into its component nutrients. To maintain soil life and organic matter levels, apply both animal and green manures and/or compost, and refrain from applying toxic chemicals.

To maintain soil condition, you must prevent erosion. This can be done in many ways, but the most important is to keep soil covered as much as possible (and especially through winter) with crops, cover crops, and mulches.

Soil with poor fertility and condition can be remediated if there are no intransigent underlying problems due to its class and/or subclass. In a small area like a garden, soil can be improved relatively rapidly and inexpensively, but on larger acreages, the job will take quite a bit more time and money.

Your Soil Improvement Program

Soil care and maintenance begin with a soil test, which involves collecting a sample and sending it to a soil lab. Soil is best sampled during the growing season, and year-over-year results will be most comparable if you sample at the same time of year and use the same lab each time you test.

Soil testing labs can be found through online searches or through your Extension office. Whichever lab you choose will give you precise directions for taking and labeling samples. Be sure to ask what the lab will test for and identify any additional tests you may want. In addition to the standard tests for pH, nitrogen, potassium, and phosphorus levels, I request organic matter percentage as well as calcium, sulfur, and magnesium levels and cation exchange capacity, which indicates a soil's nutrient balance and is indicative of its ability to release nutrients to plants.

Apply soil amendments as indicated by the test results. Your first concern should be the pH level, which should be between 6.4 and 7 for most crops. The pH level indicates the acidity or alkalinity of the soil and controls the ability of the soil and the life in it to hold and release nutrients to plants. Lime raises pH; sulfur (a component of gypsum) lowers it.

The surest way to a balanced and healthy soil is to add organic matter in the form of compost, plant matter, and animal manures — an ongoing process for the sustainable farmer. Be sure to include cover crop and green manure seed in your order, and make it a habit to plant a cover crop or apply a mulch before winter. (For a discussion of cover crops and their various effects on soil, see page 82.)

CHAPTER TWO

Late Winter

Soil surface thaws in the sunniest areas, but its average temperature is still below 40°F/4°C. Daytime air temperatures are frequently above freezing. No grass growth, though the very earliest native plants begin to emerge.

When the days begin to warm, late winter has arrived. Sap begins flowing in trees and bugs start moving around inside houses — a spider in the bed is a sure sign of winter's end at our house. It's time to wrap up woods work and remove equipment (except what you'll need for collecting sap) before the soil gets soft. Where sugar maples grow, it's sugaring time once daytime temperatures rise above freezing and nighttime temperatures stay below — this combination triggers sap flow. Traditionally this is also the season to prune fruit trees and start the first garden seeds indoors.

A fast snowmelt or a wet spring is a good time to observe how water drains across your land, where it pools, and in what order things dry out. Identifying wet and dry spots is useful for planning where to plant and graze, place culverts under trails and lanes, install drainage ditches, plant grass strips, and build earth berms to stop erosion on slopes.

Late Winter

FOCUS ON:
orchard maintenance

Garden

» **Bring out containers,** starter soil mix, lights, and other seed-starting supplies.

» **Start seeds indoors** that require the longest growing seasons.

» **Pull the mulch off** a corner of the garden, add a cold frame, and plant early greens if you're craving salad.

Field

» **Receive and store seed** away from moisture and rodents.

» **Spread manure** while the ground is still frozen *only* if there is no danger of runoff during snowmelt or spring rains.

Pasture

» **Move livestock** off wintering areas to the barnyard or to sacrifice paddock before soil gets soft.

» **Frost seed poor areas** and winter feeding area.

» **Clear brush and branches** off permanent fences and repair any broken wires and posts.

Orchard

» **Prune.**

» **Remove any tent caterpillar egg cases** you see.

» **Assess winter damage** from rodents, deer, sun, and disease.

Beeyard

» **Check for activity level** in hives (but don't open) on warm days.

» **Feed the bees.**

Barn

» **Prepare pens** for any livestock that are due to calve, lamb, or kid later this season or in early spring.

» **Vaccinate ewes or does** that are 4 to 6 weeks from lambing, introduce or give them increased grain rations, and "crutch" (shear or trim butts) them.

» **Lock in a spring purchase** of weaner pigs from a pig producer if you plan to raise weaner pigs this year.

Late Winter

Coop

» **Get the brooder ready** for chicks.

» **Clean out manure and bedding** from the coop and spread or compost it.

Equipment Shed

» **Clean and organize,** and stock up on supplies.

» **Watch for auctions and sales** of tools and used equipment.

» **Check over equipment** and start engines as time and weather allow.

Woodlot

» **Finish up timber cutting,** thinning, pruning, and burning projects for the season, in anticipation of wet and muddy conditions to come.

» **Collect and boil the sap** to make maple syrup if you have sugar maples.

Wildlife Habitat

» **Build fish cribs** (after securing a permit, if necessary) and position them on pond or lake ice so they will fall to the bottom when the ice melts.

Garden

As daytime temperatures begin to warm, you can (if you're hungry for salad) pull off the mulch from a corner of the garden and plant those peas and early greens in a cold frame, mini greenhouse, or your season-extender mechanism of choice. Peas, lettuce, spinach, onions, and parsnips will germinate in soil as cold as 35°F/2°C.

Seeds that need more clement conditions can be started inside now, or you can wait until it's warm enough to plant them directly in the garden. Which approach is best depends on a few factors: whether your growing season is long enough for that vegetable to reach maturity if sown directly in the garden, how much repotting you want to do as the plants get bigger, and whether you want to have that vegetable on your table earlier in the season. Every serious gardener I know here in the upper Midwest, including us, for example, starts tomatoes inside so we'll be eating them earlier in summer. Seed packets usually have advice on how many weeks before the average date of the last hard frost to start seeds inside, but it's advice, not a rule.

Field

Seed you ordered for spring planting should have arrived by now or will shortly; store it where it will stay dry and rodents won't get at it. If you didn't place an order earlier in the year, you can still pick up seed at a dealer's, though you may not get exactly what you want.

Before the ground thaws, you can spread manure. As soon as the soil becomes soft, keep all equipment off the fields to avoid rutting, mud, and soil compaction. You can finish spreading manure after the ground has firmed up and dried out in spring. The NOP Final Rule's stipulation about spreading fresh manure must be observed by organic farmers, but it's also a good rule of thumb for any farmer: for any crop that will be consumed by humans, manure can only be applied 120 days or more before the anticipated harvest. If you can't hit this window of opportunity, compost the manure. In general it's best to apply fresh manure to gardens only in fall.

A cold frame can give you a jump-start on the gardening season. This simple design uses just a few boards and an old window.

Starting Seeds

Starting seeds indoors does not have to be fancy or complicated: all you really need is some soil (either purchased potting soil or a home mix out of the garden and compost pile), containers to put it in (anything from egg cartons to cottage cheese containers to purchased pots or flats), a sunny window, and a relatively warm room. Plant seeds at the soil depth specified on the package, and keep the soil moist but not waterlogged. Once sprouted, put the containers in a sunny window. As plants get too big for their containers, move them into bigger pots until it's warm enough for them to be planted outside. The earlier you start seeds, the more repotting you'll have to do and the more shelf space you'll need.

You'll get much quicker and better germination if you can provide heat to raise the soil temperature (if needed) and electric light. For seeds that prefer soils warmer than your typical room temperature (and don't forget that windowsills are colder than the rest of the room), place a heating pad under the containers. In winter, when the days are still short, especially in northern regions, 12 hours a day under any type of lightbulb, in my experience, helps seedlings.

MINIMUM SOIL TEMPERATURE FOR VEGETABLE SEED GERMINATION

COLD (40°F/4°C): Lettuce, peas

COOLER (45°F/7°C): Cabbage, carrot, cauliflower, potato, radish, spinach

COOL (50°F/10°C): Beet, Swiss chard, onion, parsley, parsnip

MODERATE (60°F/16°C): Beans, cucumber, sorghum, sweet corn, tomato, turnip

WARM (65°F/18°C): Peppers

WARMER (70°F/21°C): Pumpkin, squash, watermelon

WARMEST (75°F/24°C): Eggplant, muskmelon

Note: Data compiled by J. F. Harrington, Department of Vegetable Crops, University of California-Davis.

Organic and Other Rules on Manure Spreading

Spreading manure on frozen fields is great for getting a big job done before the spring rush of work, and before the soft soil of early spring makes the task at best muddy and at worst impossible. But manure lying on top of frozen ground presents a big risk for surface waters, since snowmelt and spring rains can easily move it into lakes and rivers, where it is a pollutant instead of a fertilizer. For this reason most states regulate what degree of slope and how close to surface waters manure can be spread, especially during winter. If your fields are within a couple thousand feet of surface waters of any type, contact your Cooperative Extension Service or go online to find out what regulations apply for spreading manure.

In addition, the NOP Final Rule 205.203(c)(1) specifies that "raw" fresh, aerated, anaerobic, or "sheet composted" manures may only be applied on perennials or crops not for human consumption, or such uncomposted manures must be incorporated at least four months (120 days) before harvest of a crop for human consumption if the crop contacts the soil or soil particles (especially important for nitrate accumulators, such as spinach). If the crop for human consumption does not contact the soil or soil particles (for example, sweet corn), raw manure can be incorporated up to 90 days prior to harvest. Biosolids, sewage sludge, and other human wastes are prohibited. Septic wastes are prohibited, as well as anything containing human waste.

Organic information adapted from "Managing Manure Fertilizers in Organic Systems" by Michelle Wander, eXtension.org.

Pasture

If you've outwintered livestock on pasture, they should be moved out of winter feeding areas just before you expect the soil to thaw and/or the weather to turn wet. Keep them in the barn and barnyard or in a "sacrifice" paddock to prevent hoof action from ruining the sod when the soil gets soft. Most years we save ourselves the job of moving livestock at this time by outwintering on the planned sacrifice area and moving the cattle directly from there onto pasture at green-up. After the cattle have moved to pasture in about mid-spring, I will till the area and seed it down.

When daytime temperatures are above freezing and nighttime temperatures below, you can frost seed poor or damaged areas in the pasture. The temperature swings cause the ground to heave, working the seed into the soil. Or you can wait until the soil dries out and firms up in early to mid-spring, then plant using the more dependable techniques of drilling with a grain drill adapted for working in sod, or broadcasting (throwing) the seed and then dragging the area. Frost seeding works best with legumes (clovers) and not as well with grass seed, and your timing has to be pretty exact. With the unpredictable weather of this season, this technique has been rather hit or miss for me over the years I've tried it. But when it works, it's a quick and simple way to renovate pastures.

Make sure all of your pasture fences are in good repair, since the livestock will be moving out there shortly. Even if you've been cleaning up fence lines all winter, check them again — especially if you have trees near your fences. The big windstorms common at this time of year tend to bring down a lot of branches across fence wires.

Orchard

Pruning traditionally is done in late winter, while trees, vines, shrubs, and brambles are still dormant. When there are no leaves, it's easy to see the structure of the tree, vine, or shrub, and when the ground is a little warmer, your feet don't get so cold as you're standing there lopping and sawing. Plus, it's much easier to rake up the prunings (important for sanitation and to allow a mower to get through) before the grass starts growing.

How to Prune Fruit Trees

Pruning takes a while; if you have more than a handful of trees, you'll need to be efficient to get it all done before the buds break open. Start with a walk around the tree while you think about what you want it to look like: a tree with an open interior that allows plenty of airflow and sunlight through its well-spaced branches. This shape greatly minimizes disease problems, which are exacerbated by moisture and gloom; makes it easier for predatory birds and insects to get at the problem bugs; and encourages bigger and better-quality fruit. Prune also to the height of tree you want. Big trees require ladders to harvest; if you choose to keep trees short, the picking goes much faster.

Begin pruning by lopping off any dead or broken branches, then take off the water sprouts (the branches that have shot vertically out of bigger branches) and cut back the top to the desired height. Lastly, take out branches as needed to open up the middle. I think you often get better results more quickly by taking out a few big branches to open up the tree than by fussing with all of the little ones. We also prune to 3 feet or so above the ground, since we maintain a grass cover in the orchard and like to mow under the branches, but some orchardists let the branches droop almost to the ground.

Use sharp loppers or a small chain saw for big branches. To prevent spreading bacteria, a virus, or fungus to an uninfected tree, disinfect the loppers between trees by dipping them in isopropyl alcohol or bleach for a minute. Make cuts close to the trunk or a bigger branch, or within half an inch of a bud that faces in the direction you want the branch to grow.

The final result should be an evenly balanced tree that has well-spaced major branches with wide-angled crotches emerging from the trunk in a more or less even ladder pattern. On trees that insist differently, you can prune to have the trunk open up like fingers on a hand, so the tree looks a little like a basket. These are the best climbing trees for kids.

Late winter is the time to remove any tent caterpillar egg cases that you see.

An apple tree, pruned basket style, is a favorite for climbing.

As you prune, assess the level of damage from winter weather, deer, rodents, disease, and insects by looking for bitten-off branches, scraped trunks, holes in the bark, blackening of bark, and other abnormalities. Sometimes you can prune away the problem, but badly damaged trees should be removed and replaced.

There is disagreement on whether lateral branches should be cut back, whether all fruit trees are best pruned using these general guidelines, and various other details of pruning. Besides doing some research, it's helpful to make notes on how you pruned the trees this season, and add notes through the year on whether your technique appeared to help, harm, or not affect tree health, growth, and fruit production. I should also mention that some fruit growers are pruning during summer on the theory that this will help control overly vigorous growth and fire blight, which is a problem disease in my area.

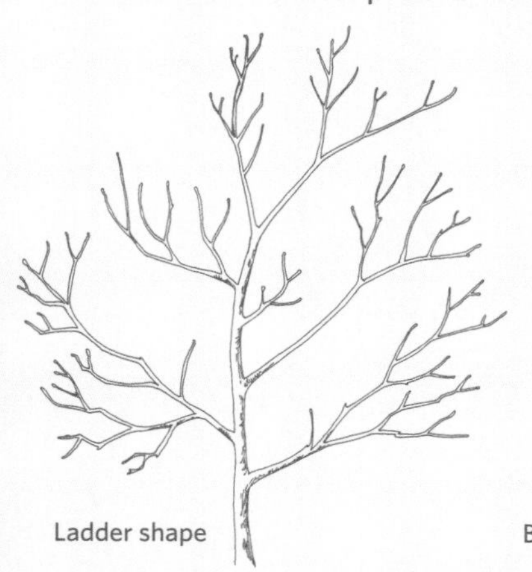

Prune fruit trees to maximize airflow and light penetration in the interior. Work with the natural shape of the tree.

Prune crossing branches on trees of all shapes.

Late Winter

Beeyard

As in midwinter, check (but don't open) the hives on warm days to see that bees are taking cleansing flights and removing dead bees. If dead bees are piling up in the hive entrance, the hive is having problems of some sort. This is a dangerous period for starvation, since honey supply is at its annual low and the bees are building numbers fast in anticipation of spring blooms. If the hive feels at all light, provide feed.

Barn

Clear out winter manure and litter as soon as possible and spread it on fields and pastures (as long as there's no danger of the manure washing into surface waters when the snow melts or in a heavy rain), but quit spreading once the ground starts getting soft. You can finish the job later when the ground firms up. Or pile the manure or compost it with old hay and straw and spread after the first cutting of hay or after crops are off the fields and garden in fall.

When to Lamb, Kid, and Calve

It's best to have your grazing livestock calve, kid, or lamb from late winter to early summer. Which date is best within this time frame depends on your situation. If the young are born in late winter, they are heavier earlier, which may translate into more ribbons at the fair or more cash in fall when it's time to sell. If you're selling direct to customers, a few weeks either way in fall probably won't make much difference to them, while a few weeks later in spring may make a lot of difference to you and to your animals.

Late winter or early spring is not an ideal time for mothers to give birth because they will have high nutritional demands but won't have any fresh green food to graze. However, babies that are born early in the season can be penned or stalled inside or close to the house, so that you can watch for problems and predators. If you plan for your livestock to birth at this time, make sure they are penned somewhere where it's dry. A pen turned mud pit will make young more susceptible to both respiratory and digestive tract infection.

Mothers who give birth later have the advantage of warmer air and ground, so babies are less likely to get cold and sick. Plus, it's much easier to get up and go out to the barn at night when it's warmer.

Clean and bed any pens and stalls you intend to use for the upcoming birthing season, and check over your supplies. Sheep and goats are commonly put into individual pens made with movable panels for ease of dealing with any delivery problems and to ensure good bonding between mother and youngster.

Pregnant ewes and does require special treatment at this season, since 70 percent of fetal growth in these species occurs in the last several weeks of pregnancy. Four to six weeks before lambing or kidding, gradually add or increase the grain in the ewe's or doe's daily ration, with the goal of having them be neither too thin nor too fat on their due dates to minimize delivery problems. Adjust the amount of grain according to the age of the ewe or doe and the number of young she is carrying, if known from an ultrasound. In general, first-time mothers need more grain than older ones, since they are still growing, and mothers carrying multiple fetuses need a richer diet as well. Many shepherds sort their ewes into separate feeding groups based on their various needs. If the hay quality is low (grassy or old), consider feeding supplemental calcium. Give annual vaccinations two to four weeks before due dates. Deworm if indicated. Lastly, "crutch" the ewes and does (shear the area around the udder and vulva).

Cattle generally don't need any special preparations for calving if their body condition is good and their vaccinations are current. Heifers that are still growing while pregnant might benefit from grain in their ration.

If you're planning to raise pigs this year, now is a good time to arrange a date to pick up weaner pigs from a pig producer. The pigs will reach butchering weight in late fall, so you can put a year's worth of pork in the freezer without having to keep boars and sows yourself.

Coop

As the days warm, hens will be spending more time outside, so you can get inside the coop and start shoveling out winter's accumulation of bedding and manure without traumatizing the birds too much. It's also still cool enough to make this dusty, smelly job at least tolerable.

Equipment Shed

This relatively slow time of year is an opportunity to clean and organize tools and supplies and take stock of what you have and what you're going to need for supplies and parts for the coming growing season. On warmer days, start all the engines one by one to make sure they will run smoothly when you need them.

Late winter and early spring are busy times for auctions and equipment sales, since sellers know they get better crowds this time of year, when folks are a little restless as winter is winding down. Watch the sales notices if you're looking for a specific tool or piece of used equipment, and be sure to take a knowledgeable friend along to check over anything you're interested in buying.

Woodlot

As the big thaw begins, wrap up woods projects for the year and move out equipment, firewood, and logs before the ground turns soft. The big storms common in late winter can bring down branches and trees across trails; these should be picked up to clear the way for summer mowing. Stack the wood for next winter's fires, or add it to brush piles for wildlife habitat.

When daytime temperatures start rising regularly above freezing but nights do not (ideally no warmer than 40°F/4°C or colder than 24°F/–4°C), it's prime maple sugaring season. Maple syrup is made from the sap of sugar maples, also called hard or rock maples. If you don't have loamy soil or the right climate for sugar maples, you can make syrup from any type of maple, as well as from birches, walnuts, butternuts, and probably several other types of trees. It will just take a lot more sap to get the equivalent amount of syrup. You can use traditional spiles and buckets or plastic hoses and sacks.

Traditional spiles and buckets work fine for a small sugaring operation.

Tick Season Begins

Ticks start showing up on clothes, skin, and pets in our area as soon as melting snow reveals bare ground. Besides being an irritation, ticks can carry a variety of potentially debilitating diseases, including Lyme, erlichiosis, and babesiosis, depending on where you are in the country.

You learn to live with ticks. During tick season, we wear knee-high rubber boots in the woods, tucking our pants inside. Keeping grass on the trails short keeps the tick load there to a minimum; if you let your laying hens run free in the yard, they will eat the ticks. We pour a tickicide on the dogs once a month and check ourselves every evening. Showering at night instead of morning can wash off hard-to-spot tiny ticks before they attach, and we also wash our clothes daily during tick season. We keep a "tick bowl," a small bowl filled about two-thirds full with water and a few drops of dish detergent, at hand in the pantry. Any tick we find is dropped in the bowl, and the tick sinks to the bottom (thanks to the detergent cutting the surface tension of the water) and drowns. If you suspect you're becoming ill due to a tick bite, see a doctor as soon as possible — the quicker treatment is started for a tick-borne disease, the better your chances are that it won't become serious.

Wildlife Habitat

Fish cribs protect small fish from being preyed upon by large fish and other predators. Designs vary; most are four-square squat towers of logs stacked so the sides are semi-open. The interiors should be filled with woody brush for more protection. Cribs are built on or towed onto frozen lakes and ponds, where they'll drop into place when the ice melts.

Most states regulate the installation of fish cribs, so you may need a permit. Contact your state Department of Natural Resources for information and technical assistance.

Topic of the Season

Meeting Plant Needs

The diverse plants that clothe our acres all share the same basic needs for water, sunshine, fertile soil, and space. But each plant varies in its preferences and in what it gives back to the land. The various maples, for example, tolerate shade well as young trees, but most garden vegetables thrive only in full sun. Root crops do best in light soils, while fruit trees prefer loam and demand good drainage. Red clover will grow on wetter and heavier ground than alfalfa will tolerate. The legume family — beans, peas, clovers, alfalfa, and some woody plants — fix nitrogen in the soil. Grasses are amazing for holding soil in place and generous in contributing organic matter in the form of constantly growing and dying roots. Crops like corn and melons are greedy users of soil fertility but deliver high amounts of human and livestock food in proportion to the space they occupy.

In this section, I'll discuss annual plants. Perennials are handled differently: plant species are not rotated and soil improvement amendments are used less frequently and are not incorporated into the soil. I will cover how to maintain soil health and fertility with the most common perennial crops — pasture, permanent hay fields, fruit trees, and timber — in chapters 6, 7, and 10 (pages 97, 112, and 154).

Crop Rotation

Good farmers have known for millennia that growing the same crop on the same ground every year increases disease and pests, depletes the soil of fertility, and reduces the harvest. Though it has been forgotten by many conventional farmers in the present era of synthetic fertilizers and pesticides, crop rotation is fundamental to sustainability.

IN THE FIELD. The classic American system, especially well adapted to New England, the Mid-Atlantic states, and the Midwestern states, was established sometime after the Revolutionary War. This is a four-year rotation of corn, small grain (fall-planted winter

Keep Off the Soil!

As frost leaves the ground from late winter through early spring, soil is at its softest and most vulnerable. Move equipment, logs, and firewood out of the woods before this happens, and avoid running any equipment over newly thawed bare ground. Don't till in fields or garden until the soil has firmed and dried. Tillage and traffic at this time will not only get you stuck in the mud, they will compact the soil, destroying structure and making it harder for air, water, and roots to penetrate. Compacted soil is the reason why so few things ever grow in wheel ruts on trails.

Crop rotation is key to pest and weed control, as well as soil health.

wheat or spring-planted oats), and two years of a legume hay — originally clover and later alfalfa. If you are raising livestock so that the hay is fed on the farm and the manure returned to the soil, this is a fine basic system that's good for soil health: it prevents the buildup of crop-specific pests and diseases and spreads field work more evenly through the growing season.

Corn demands fertile soil, and since it's planted in rows, you can have good weed control. Small grain, which follows, benefits from the previous year's weed control and needs less fertility, requires less attention, disrupts the life cycles of many insects and diseases that thrive on corn, and is planted and harvested earlier than corn. You may plant a hay crop of legumes with the spring grain or plant a mixed hay of legumes and grasses after a winter grain has been harvested (I've also planted oats as a nurse crop for alfalfa in late summer, which is too late to get a grain harvest but worked beautifully for sheltering the young alfalfa from winter temperature swings). Forages not only feed livestock but build soils, adding organic matter and preventing erosion, and are harvested primarily in summer, when the farmer is not busy with corn or small grains. The NOP Final Rule specifies that at least three different crops must be planted over five growing seasons, and these must include sod crops, green manures, and cover crops.

The traditional American field crop rotation works well in the Corn Belt and the Northeast, but not so much in the dry West or hot and humid Southeast, or if you don't need hay for livestock. In the dry Great Plains states, traditional rotations are built around small grains. In the hot and humid Southeast, peanuts may replace corn in the rotation. Where there are no livestock to process forages, green manures can be planted and tilled in to replace animal manure. Information on sustainable crop rotations for different regions is available from your state Cooperative Extension Service, beginning farmer programs, and ATTRA (Appropriate Technology Transfer for Rural Areas), a division of the nonprofit National Center for Appropriate Technology (NCAT).

IN THE GARDEN. Plants are rotated by family; I aim for three years or more before planting the same thing in the same place. For example, potatoes, tomatoes, peppers, and eggplants all belong to the nightshade family (Solanaceae), so once any of these are grown in a particular bed or row, no other plants in the family should be planted there for three years. The effect of rotating garden crops is the same as for field crops: fewer disease and pest problems, and more balanced demands on soil fertility.

In your early years of growing, planning a good rotation that includes not only crop plants but also green manures, trap crops, plants for beneficial insects, and cover crops can take some time. After several years' time, when experience has bred familiarity with the various plant families and their effects on the soil, rotations become easier to plan and more intricate as you fine-tune your system.

Light and Heavy Feeders

In addition to considering your crops' effects on soil health, pest and disease cycles, and your need for income when building your crop rotation, think about your crops' fertility needs. "Heavy feeders" demand high soil fertility, "light feeders" need only moderate fertility, and soil-building crops help restore both fertility and structure. It makes sense to plant these three categories of crops in that order, in both field and garden. The traditional corn-small grain-forage rotation does exactly this, but there are many options, and you should adapt your rotation to what you want to raise.

Heavy feeders: Asparagus, beet, broccoli, Brussels sprouts, cabbage, cantaloupe, cauliflower, celery, collard, corn (field and sweet), eggplant, endive, kale, kohlrabi, lettuce, okra, parsley, pepper, potato, pumpkin, radish, rhubarb, soybean, spinach, squashes (summer and winter), strawberry, sunflower, tomato, watermelon

Light feeders: Carrot, garlic, small grains, leek, mustard greens, onion, parsnip, rutabaga, shallot, sweet potato, Swiss chard

Fertility builders: Legumes (all peas and beans, clovers, alfalfa, peanuts)

Note: List adapted primarily from the *Arizona Master Gardener Manual* produced by the Cooperative Extension, College of Agriculture, the University of Arizona.

The USDA Organic Rule on Crop Rotations

National Organic Program (NOP) Final Rule 205.205 defines crop rotation as "alternating annual crops grown on a specific field in a planned pattern or sequence in successive crop years so that crops of the same species or family are not grown repeatedly without interruption on the same field. Perennial cropping systems employ means such as alley cropping, intercropping, and hedgerows to introduce biological diversity in lieu of crop rotation."

The producer must implement a crop rotation including but not limited to sod, cover crops, green manure crops, and trap crops that provide the following functions that are applicable to the operation:

- Maintain or improve soil organic matter content
- Provide for pest management in annual and perennial crops
- Manage deficient or excess plant nutrients
- Provide erosion control

The NOP standard requires that all functions applicable to the operation be addressed. Criteria to judge compliance with the standard are subjective. No technical measurement tools are required to evaluate the effectiveness of a rotation for building organic matter or controlling erosion.

Certifying agencies look for:

- At least three different crops planted over five consecutive years in annual cropping systems
- A leguminous green manure crop planted at least one time in the 5-year period to improve soil
- Diversity in plant species or families so that crop species or crops within the same family are not grown repeatedly without interruption on the same field
- Inclusion of crops with different rooting systems and pest pressures to assist with building soil quality and reduced pest management activities
- Use of green manure and cover crops
- Crop rotations (as opposed to a double-crop system; for example, growing a continuous vegetable crop with a wheat cover crop)

Note: Adapted from the NRCS Practice Fact Sheet "Organic Production: Using NRCS Practice Standards to Support Organic Growers," April 2010.

Plant Families

Amaranthaceae: Amaranth

Amaryllicaceae (onion and lily family): Chives, garlic, leeks, onions, shallots

Apiaceae or Umbelliferae (carrot family): Carrots, celeriac, celery, chervil, coriander, dill, fennel, parsley, parsnips

Asteraceae or Compositae (sunflower or aster family): Artichokes, endive, Jerusalem artichokes, lettuce, sunflowers

Brassicaceae or Cruciferae (mustard family): Asian greens, broccoli, Brussels sprouts, cabbage, cauliflower, collards, horseradish, kale, kohlrabi, mustards, radishes, rutabagas, turnips, watercress

Chenopodiaceae (beet or goosefoot family): Beets, quinoa, spinach, sugar beets, Swiss chard

Convolvulaceae (bindweed/morning glory family): Sweet potatoes

Cucurbitaceae (squash, cucumber, and gourd family): Cucumbers, gourds, melons, squash

Fabaceae or Leguminosae (bean or legume family): Alfalfa, beans, clovers, lentils, peanuts, peas, soybeans, sweet clovers, vetches

Malvaceae (mallow family): Okra

Poaceae or Gramineae (grass family): barley, broom corn, corn (dent, flint, flour, popcorn, and sweet corn), millet, oats, rice, rye, sorghum, wheat

Polygonaceae: Buckwheat, rhubarb

Rosaceae (rose family): Almonds, apples, apricots, blackberries, cherries, peaches, pears, plums, quinces, raspberries, strawberries

Solanaceae (nightshade family): Eggplant, ground cherries, huckleberries, peppers, potatoes, tomatoes

Information adapted from *Seed to Seed: Seed Saving and Growing Techniques for Vegetable Gardeners*, by Suzanne Ashworth (Seed Savers Exchange, 2002).

CHAPTER THREE

Early Spring

Soil thaws and warms through the season to 40°F/4°C and higher. Cool-season grasses begin to grow, but slowly. Spring small grains are planted as soon as the ground can be worked.

MUD, FLOOD, BLOSSOMS, and babies signal the arrival of spring, though the surest sign is the frost going out of the ground — and staying out. In early spring the work that's needed during the growing season really starts: at first it's just a trickle, but within a couple of weeks it feels like trying to drink from a fire hose. Unless it is dry, the soil is at its softest just after the frost goes, making this the easiest time of year to put fence posts in the ground. As the soil firms and dries, tilling and planting get under way and livestock can go out on pasture (but keep feeding them hay). Traditionally it's also rock-picking time, if you have rocky soil.

Early Spring

FOCUS ON:
fields • garden • fences • barn

Garden

- » **Plant peas** and other cold-tolerant vegetables as soon as the soil softens.
- » **Spread compost** and purchased amendments (such as lime and blood meal) on the rest of the garden when the soil begins to firm up.
- » **Till as soon as the soil is dry** and firm enough to work.
- » **Start more seed inside** for mid- and late-spring plantings.

Field

- » **Check for winterkill** in the hayfields as green-up begins.
- » **Reseed any affected areas** when the soil can be worked and at the same time that you disc and overseed winter feeding areas.
- » **Spread manure** (from winter feeding areas and the barn) in the crop fields as soon as the soil is dry enough to run equipment without rutting or getting stuck in the mud, then till under both manure and winter cover crops.
- » **Pick up rocks.**
- » **Finish tillage,** such as discing or dragging, to prepare a good seedbed.
- » **Plant spring small grains** as soon as possible.

Pasture

- **Finish clearing and repairing** permanent fences.
- **Start putting up temporary fencing** for rotational grazing.
- **Disc and seed poor areas** if they haven't been frost seeded.
- **Start grazing** as soon as the ground is dry enough that hooves won't cause serious damage.

Orchard

- **Finish pruning** before green-tip bud stage, remove the debris, and decide whether to save or replace damaged trees.
- **Plant new trees.**
- **Continue to look for and destroy** tent caterpillar egg cases and nests.

Beeyard

- **Continue to feed** if indicated up until the first major nectar flow.
- **Open hives** on a warm day to check for health and strength.
- **Medicate** according to regional recommendations.
- **Clean up** any winter mouse damage.
- **Replace worn combs** and parts.
- **Place new hives,** and hive new queens and bees.

Barn

» **Finish winter manure cleanup.**

» **Finish spreading** or stack or compost the manure.

» **Keep livestock penned** in the barnyard or sacrifice area until the soil dries and firms up.

» **Calving, kidding, or lambing** begins for most.

» **Prepare sty** or hut and paddock for weaner pigs, and lock in an agreement for picking them up after weaning.

Coop

» **Set up the brooder** and get the temperature right several days before new chicks are due to arrive.

» **Purchase chick starter feed** and chick-sized grit ahead of time.

» **Install new chicks** in the brooder.

Equipment Shed

» **Pull out tillage and planting implements** so they're ready to go.

Woodlot

» **Plant trees** and control invasive species.

» **Maple sugaring ends** as both night and daytime temperatures rise.

» **Seed patches of raw dirt** on the trails and spread a light mulch.

Wildlife Habitat

» **Till and plant native grasses** and forbs in set-aside areas as soon as the soil can be worked.

» **Put up birdhouses** built during winter.

» **Burn established prairies,** if needed, at this season.

Garden

As soon as the soil is dry enough to walk on, remove any mulch, apply any amendments, then till or dig the garden, turning under any winter cover crops. (A heavy mulch or cover crop may not till under easily and may have to be mowed first, or removed and composted instead.) If you are observing the NOP Final Rule, don't put fresh manure on the garden unless you are certain there will be at least 120 days before harvest. Better yet, don't put any manure on the garden in spring; we apply it on the hay fields in spring and manure the garden in fall. Compost can be spread this season, however, if you didn't spread it last fall and your compost pile has thawed.

Cold-tolerant vegetables can be sown in the garden; continue to start cold-intolerant and long-growing-season vegetables indoors or outdoors in cold frames or other protective structures (see page 26 for information on starting seeds indoors and soil temperatures for germinating different vegetables).

Order of Planting

Plants differ not only in their tolerance of cold and heat, they also take very different amounts of time to sprout, grow, and mature. Vegetables and herbs that need a long growing season and don't tolerate cold should be started indoors; those that like cooler conditions and mature rapidly can be planted directly in the garden as soon as the soil is warm enough for that type of seed.

The basic planting schedule below is built around the average date of the last frost. You can vary it according to packet directions for a particular variety (there can be considerable variation within a species) and your own inclinations.

PLANT TYPE	START INSIDE	TRANSPLANT TO GARDEN	SOW DIRECTLY IN GARDEN
Slow to sprout Parsley, leeks, onions (seed)	15–20 weeks before last frost		
Very hardy Peas, lettuce, crucifers (or plant midsummer for fall harvest), spinach, parsnip, spring small grains	10–14 weeks before last frost	7–9 weeks before last frost	As soon as soil is thawed and you can erect a mini hoop house or cold frame
Moderately hardy Potatoes, leeks, onion sets, root crops (or plant midsummer for fall harvest), field corn			4–6 weeks before last frost
Tender Tomatoes, peppers, sweet corn, tender herbs, beans, squash and melons	4–6 weeks before last frost	After last frost	After last frost and after soil is warm

Information adapted from *The Week-by-Week Vegetable Gardener's Handbook*, by Ron Kujawski and Jennifer Kujawski (Storey Publishing, 2010).

Field

As the hayfields green up in spring but while it's still too wet to work the soil, walk your fields to check for winterkill (damage from soil heaving, which occurs when nights are below freezing and days are above; alfalfa is especially susceptible) in the legumes. If there are poor areas, disc and replant them at the same time you're discing and overseeding winter feeding areas.

As with the garden, tillage and seedbed preparation in the field begins as soon as the soil has dried and firmed. After any manure is spread, it is tilled under along with any cover crops, then the seedbed is smoothed. Plowing is the most effective way of incorporating manure, green manure, and amendments into the soil, but you may want to research alternatives, ranging from fall plowing before planting a winter cover crop and discing it in the following spring to no-till systems for your acres. Many sustainable farmers have found that using tillage implements (a chisel plow is one type) that stir the soil without burying crop residue or winter-killed cover crops will better hold moisture in the soil, improving germination in a dry spring. To control weeds, many organic farmers use a shorter variation of the "stale bed" method, cultivating their fields once or twice immediately before planting to kill any emerging weeds (most of the young weeds won't even be visible).

If you have field rocks, pick them up after discing and you'll have fewer equipment repairs.

Small grains are planted when tillage is complete; most of the row crops need warmer soil and so those aren't planted until mid-spring. If you plant too early, when the soil is too cold for your particular type of seed, the seed may rot in the ground and germination may be uneven; early planting may also give weeds a head start if they sprout before the crop does. But if you plant too late in the season, the crop may not have enough growing days to reach maturity, reducing the yield. Fungicide-coated seed may resist rotting, though it is not allowed in an organic system. The biggest difficulty with timing planting correctly is the highly variable weather of spring — and if it's a cold, wet one, it's difficult to get a crop in on time without risking rutting and getting stuck (believe me, I've been there).

Recommended Minimum Soil Temperatures Needed for Germination of Field Crops

Crop	Temperature	Crop	Temperature
Spring wheat	37°F/3°C	**Soybean**	59°F/15°C
Spring barley	40°F/4°C	**Sunflower**	60°F/16°C
Rye	41°F/5°C	**Millet**	60°F/16°C
Oats	43°F/6°C	**Sorghum**	65°F/18°C
Field corn	55°F/13°C	**Dry bean**	70°F/21°C

Note: Adapted from "G96-1362 Soil Temperatures and Spring Planting Dates," by Stephen Meyer and Allen Dutcher, University of Nebraska–Lincoln Extension.

Time and Labor Needed to Perform Field Work

Plowing, discing, dragging, planting, cultivating, and harvesting take considerable time. Just how much depends mostly on the size of your equipment, but as a reference point, Karl Schwenke in his excellent *Small-Scale Farming: An Organic Approach* (Storey, 1979 and 1991) estimates that to plant, cut, shock, and husk by hand 10 acres of field corn requires approximately 240 hours of labor from start to finish.

I know it takes me about 12 hours to cut 30 acres of hay with a 9-foot haybine (mower) hitched to our 50-horsepower tractor if the crop isn't too heavy, plus about 4 hours to rake it with the V-rake (it took 7 hours or better with the old side rake), and about 4 hours to bale it with our round baler. With the old square baler, it used to take 12 hours to bale, as long as nothing broke. None of these estimates include the additional time it takes to get equipment out of the shed and hitched up, and to make repairs when something breaks, which usually happens and you hope to heaven it's a minor problem.

Growing Small Grains

Small grains can be used in field or garden as cover crops or green manures, or for livestock feed, straw, or human food. In general they'll tolerate less-fertile soil and need less care than field corn.

There are two types:

Winter grains are planted a few weeks before the first expected hard frost in fall and go dormant through winter (precise planting times vary by the type of grain). They green up in early spring and are ready for harvest by early to midsummer. Winter grains include winter wheat, spelt, winter rye, and winter barley.

Spring grains (except for millet and amaranth, which need warm soil) can be planted as soon as the ground can be worked in spring (harvest season varies by type). Spring grains can also be planted in fall as a winter cover crop; the plants will be killed by the frost before they can produce grain, so they will function as a mulch through winter. Types include spring wheat, spring barley, millet, oats, buckwheat, amaranth, and quinoa.

Pasture

Finish inspecting and repairing permanent fences, and pick up any sticks and branches that have fallen on the field or pasture side of the fence so that you'll be able to mow later. As in the orchard, do this before the grass starts to grow, while you can still see the sticks on the ground. Also fill in any big holes that have been dug by critters, which are a danger to equipment and to livestock that could twist or break a leg from stepping into them. This is the best time of year to build new permanent fences or to replace bad ones, since the soil is so soft that the posts just about sink in by themselves.

As soon as the ground has firmed up enough, I turn the livestock out on the pasture. They're anxious to get out and graze at this time and will find food where, to our human eyes, there's nothing to eat. Continue to feed hay until your animals quit eating it, and you'll have a smooth transition from winter to summer diets.

We have an 8-acre fenced field right next to the house that I use for both hay and pasture. Our cattle go in there as soon as the soil is firm, and they calve where I can keep an eye on them from the kitchen window. This puts that field a week or two behind our other pastures and hayfields, setting the grazing rotation in motion. If I decide to cut it for hay, it will be cut after the rest is done, spreading out the workload.

Just before and then as livestock begin to graze, start building paddocks for rotational grazing. If you birthed on grass, you can gradually subdivide that pasture and drift the flock or herd into smaller areas as grass growth accelerates, and the babies will become accustomed to the electric fence. (For more on rotational grazing and electric fencing, see page 66.)

Livestock will begin grazing before you can see anything for them to eat. Keep feeding hay until the animals quit eating it.

Orchard

Tent caterpillars can quickly strip a tree of leaves. Now is the time to control them, before the leaves obscure their nests. Tear the nests open with a gloved hand or a stick and scatter the worms for the birds to eat.

If you haven't finished the late-winter pruning, get it done now so you don't open wounds on trees when airborne pathogens start to move again. Pick up and remove all trimmings (especially branches that look diseased or have signs of insect infestation) both for sanitation and so you can move a mower through the orchard. Chip, burn, or pile trimmings well away from the orchard. Cut down and replace any dying or dead trees, which show no green under the bark.

New trees are traditionally planted as soon as possible after the ground has thawed, before they break dormancy. Be sure to place hardware cloth around the base of the trunk to protect them, and set up an electric deer fence or a tall cylinder of woven wire. Wild animals are hungry this time of year, when winter food supplies are depleted and tender green saplings look very tasty. (For a discussion of how to choose varieties and site, and planting fruit trees, see page 112.)

Tent caterpillar nests are easiest to see and destroy before the trees leaf out.

Beeyard

Hive activity should be increasing as the days get warmer and flowers start to blossom. Do the spring cleaning chores now: the bees will make more honey during major nectar flows later if you don't disturb them with cleaning at that time. There is still danger of starvation until bloom season really gets under way in mid-spring, so continue to feed until the bees are no longer interested in what you offer. The hives can be opened on a warm day to check bee health: bees should be numerous and busy, with plenty of brood cells. Medicate according to the recommendations of regional honey bee associations, clean up any damage made by mice in the winter, and replace any combs and parts that are worn. New hives should be placed in the field, preferably where they will get morning sun, receive light afternoon shade in summer, and be protected from high winds, livestock, equipment traffic, and agricultural chemicals. Hive new queens and bees when they arrive, according to the instructions on the package or recommendations from other local beekeepers.

Barn

Kidding, lambing, and calving seasons are beginning or soon will, so make sure you have on hand any supplies you might need, including frozen colostrum, drying towels, heat lamps in cold weather, iodine for dipping navels, a rectal thermometer, and medications and vaccinations appropriate to the species. Post your veterinarian's phone number next to the phone, and ask her ahead of time for regional recommendations for the type and timing of vaccinations and other disease-preventing measures. Then it's mostly watching and waiting.

Animals about to give birth may start acting uncomfortable several hours before the event, or they may not show many signs at all. When (or if) you see a rush of liquid from the vulva, this means the water has broken and the baby should follow within an hour or two. If it doesn't, pen the animal when possible (this can be tough with beef cows). If no nose or hooves appear within an hour or so, or if they appear and make no further progress, consult your vet or a knowledgeable friend or neighbor. Many longtime livestock owners have become quite expert over the years at rearranging legs and heads inside the uterus so a birth can proceed. It is also well worth your time to read up on how to recognize a normal (nose and front feet first and together) versus an abnormal birthing process for the species you own, and what you can do and when you should get help.

Once a youngster is on its feet and nursing, you can relax. But continue to keep an eye on things — infections can develop rapidly in young stock; if you observe diarrhea, failure to nurse, unwillingness to move, or other abnormal behavior in a kid, lamb, or calf, then pen the animal and take its temperature with a rectal thermometer before contacting your vet, so you can let the vet know its temperature. The mothers, too, are subject to various problems and should also be monitored, though most of the time if you've paid attention to diet and mineral supplements and general health and comfort through the pregnancy, there will be few if any problems.

Protecting Flocks from Predators

Although small livestock are vulnerable to predation at any time, they are never more so than during lambing and kidding. Even cattle, who are generally capable of defending themselves against almost all comers, lose calves to determined predators. Chickens are vulnerable at all times.

Livestock owners can employ various lines of defense, including appropriately configured and powered electric fences, guard animals that have been raised with the livestock, and keeping animals inside during vulnerable periods. Chickens especially should be cooped up at night for their own protection throughout their lives, whether inside a permanent coop or a temporary mobile structure out in the pasture. They will come inside on their own each night; your job is to make sure the door is shut and latched after they're tucked in for the evening.

Coop

If you will be starting new chicks, it's time to pick up your chick feed and get the brooder ready, before they're due to arrive. Turn on the heat lamp in the brooder and place a thermometer under it for a couple of days, until the floor has warmed up and the temperature equilibrates. Then adjust the lamp up or down until the temperature directly under it is steadily between 90 and 95°F/32 and 35°C.

You can start chicks in winter, but waiting until early spring means it's easier to keep them warm, and when they're big enough to go outside there will be green grass, bugs, and sunshine. If you keep feathered-out young poults cooped up longer than necessary because it's still too cold and wet, they may take out their frustrations by bullying smaller birds, pecking them and keeping them from food and water, and you'll have more chicken manure to haul. It's easier to get your chicks later in the season! (See page 14 for more on how to time the ordering and raising of both meat and laying birds.)

New chicks need a secure, draft-free brooder with a constant temperature, clean water, fine grit, and starter feed.

When your chicks arrive, open the box inside the brooder and lift them out one by one, using your thumb to push each of their beaks into the water to get their first drink before you turn them loose. Check them several times that first day to make sure they've found the feed and water and the warmth of the lamp. Don't be concerned if your chicks have all flopped down with their wings out, looking as if they're dead — they're likely sleeping. Adjust the height of the lamp if needed. If they rest in an even circle under the lamp, it's just right. If they're crowded into the center or piled in the corners, it's too cold or too hot, respectively. Keep feeders and waterers clean and full. Sprinkle grit on the feed for a week or two before feeding separately, and throw in green grass and even a chunk of sod every day to keep them healthy.

Equipment Shed

Pull out and check over equipment in the order that you will need it: spreader, plow, tiller, disc, drag, planters, cultivators, and mowers. Make sure you have the necessary hitch pins, the tires are fully inflated, and all moving parts move. Any repairs or adjustments you can get done now save time later, when you will really need to save time. Have a supply of fuel on hand (we just fill a bunch of gas cans). As you finish with each implement and as time allows, clean it up, check it over, and tuck it back into storage.

Woodlot

Though logging and other woods projects that require vehicles and equipment are on hold at this season, it's the perfect time to tackle two jobs in the woods: pulling up invasive plants and planting new trees. The soft soil of early spring makes this the best time of year for uprooting plants — especially the woody, brushy ones that hang on so tightly when the soil is dry. Small specimens can be pulled by hand; use a mechanical puller or an electric winch on larger plants. Pile the woody debris for burning next winter, and compost the green plants.

Trees can be planted as soon as the ground is frost-free. Trees for woodlots are generally bare-root, dormant one- or two-year-old seedlings ordered by the dozens or hundreds. They should be put in the ground as fast as possible before the roots can dry out, so the technique tends to be a little more rough-and-ready than when you're planting orchard trees. (See page 154 for more on how to plant trees for woods.)

After you're all done with all of these projects, repair ruts and seed the bare parts of trails.

Cut off and burn the tops of invasive woody species during the winter, then return when the ground thaws in early spring and pull the stumps.

Wildlife Habitat

Where you've planned to plant native prairies or other mixes of native grasses and forbs, do final seedbed preparation as soon as the soil can be worked, then broadcast seed.

Birdhouses and bee and bat houses built over the winter should be installed early, before migrants return and insects emerge. Consult guides for the best heights and habitats to place these in.

Topic of the Season

The Lives of Poultry and Livestock

Animals complete the ecology of the homestead, processing plants not usable by humans into high-quality food and enriching the soil with their manure. The forages we grow to feed our livestock are great soil builders as well. Pigs and chickens are naturally omnivorous garbage disposals, turning all manner of scraps and discards into meat, eggs, and high-quality manure. Grazing animals spread their own manure as they graze and have a symbiotic relationship with forages. Mimic their natural grazing patterns with rotational grazing (discussed in detail on page 66), and your pastures and the soils under them will thrive. In fact, grazing livestock when properly managed are one of the finest means available for improving soil health and preventing erosion.

As with plants, all animals share the need for clean and adequate water, a diet appropriate to their species, a clean environment, fresh air and sunshine, shelter from bad weather, and attention to their natural living habits.

Poultry

Chickens (and occasionally ducks and other fowl) are raised for meat or to produce eggs, and some breeds are good for either purpose. Meat breeds gain weight quickly and produce larger carcasses, while laying breeds produce more eggs. To get started with poultry, you can order eggs or chicks from a hatchery or buy some directly from a grower. I like to order my chicks for early spring, when it's warm enough that I can set up a brooder in an unheated outbuilding instead of the house, because cute as they are, those tiny chicks are smelly and dusty! By the time they're feathered out at about six weeks and ready to be on pasture

Sheep are well suited to small acreages.

in portable pens, the grass is just right for them — not too tall or too short.

Poultry require grit for digestion, since they have no teeth, and a high-energy diet for good production (either your own grain mix or a commercial feed). Commercial chicken rations come in starter, grower, and layer formulations. Switch to grower when chicks are feathered out, and to layer a couple of weeks ahead of when the birds are expected to start laying. Geese prefer more grass in their diet than other poultry.

Geese and turkeys are generally butchered in fall after a spring hatching, after they've started to put on fat. Ducks and chickens are butchered at variable times, depending on how large a carcass you want. The Cornish-Rock chicken is the best breed for fast growth; it is ready to butcher at 8 to 10 weeks. Birds can be butchered at 3½ to 5 pounds live weight for broilers or at 6 to 8 pounds for roasters. Older poultry (meat or egg birds) can be butchered and eaten as well; they're just tougher and so best in soups and stews. For laying hens, the most productive breed is the Leghorn, but I've always preferred the more durable and docile heritage breeds.

Pigs

The fastest growing of all domestic farm animals, pigs are born weighing about 3 pounds and will reach 250 pounds or more in six months. Male pigs being raised for meat should be castrated to avoid "boar taint" in the meat and to make the animal less aggressive.

Heritage Breeds

When choosing a breed of poultry or livestock to raise, consider heritage breeds: those animals that have been bred for durability and adaptability to a particular environment. Large conventional livestock operations focus on breeds that produce the most milk, eggs, or meat. Any other considerations often go by the wayside, such as an animal's ability to produce while staying healthy, living long, utilizing the food sources native to the region, and tolerating the weather.

Heritage breeds are more durable, more adaptable, and often more docile than conventional breeds, making them especially suited to small acreages and living outdoors. Check out the breed descriptions at the Livestock Conservancy website (see Resources).

Poultry Life Cycles

Note: These figures are averages; they will vary by breed and weather.

	CHICKEN	DUCK	GOOSE	TURKEY
Incubation	21 days	28 days	29–31 days	28 days
Fully feathered	6 weeks	6–7 weeks	6–7 weeks	8 weeks
Egg laying begins	5–7 months	About 6 months	In spring	The spring after hatching

Swine Life Cycle

Gestation period: 112 to 115 days (3 months, 3 weeks, and 3 days)

Litter size: Average 8 to 12 piglets per litter

Castration: At 4 to 14 days, no later than 5 days before weaning

Vaccination: For atrophic rhinitis before weaning; for erysipelas when purchased as feeder pigs

Weaning: At 6 to 8 weeks traditionally (commercial pig farms wean at 3 to 4 weeks)

Butchering: Traditionally in late fall at 5 or 6 months and a live weight of anywhere between 200 and 300 pounds

Sexual maturity: At 5 to 6 months for both sexes

Rebreeding: Can occur within a week after weaning

Cycle: Pigs cycle year-round every 17 to 21 days.

Litters per year: Two on average, with a not-quite-4-month pregnancy, then a 6- to 8-week lactation

Life span: Highly variable. A sow may produce piglets for 4 to 5 years if the owner isn't too worried about maximizing piglet production; on commercial farms, sows tend to be culled much earlier.

Diet: Because they grow so fast, pigs demand a nutrient-dense diet. Pigs have traditionally been raised on dairy waste, table scraps, garden leftovers, and other garbage, but many small producers now will mix commercial feed with waste for a more complete diet. Pigs will eat some grass, and they will root for insects, roots, and small animals.

Water: Fresh water should be available at all times; pigs also like a mud puddle to wallow in on hot days.

Shelter and outdoor access: Pigs love to be outside, but they don't tolerate either extreme heat or cold and so should always have access to a draft-free hut or sty. They should have plenty of bedding in winter to stay warm.

Sustainable, humane rearing: Pigs are happy on pasture during warm seasons, as long as they have shelter from bad weather. Use them as natural plows to clear and prepare ground for planting. *Young* pigs can be run through the orchard at apple drop time to clean up the drops.

Other considerations: Pigs are tough to fence in with nonelectric fencing, but they are very sensitive to a strand or two of electric wire at nose level. Check the wire frequently, since their rooting activity can throw soil on the fence and cause it to short out.

Sheep, Goats, and Cattle

Cattle, sheep, and goats are raised for both meat and milk and share the characteristic of being able to survive on forages — grasses and clovers. But they grow faster and bigger, and produce more milk, when nutrient-dense foods like grain are added to their diets. However, too much grain or similar foods, especially if there's not enough exercise or forage in their diet, makes for unhealthy animals with shorter lives. As with pigs and poultry, grazing livestock breeds vary widely in size, appearance, and growth rates. In general, goats don't tolerate cold or wet weather well, and sheep don't do well in wet, warm climates. Wool sheep are sheared annually in late spring or early summer.

Grazing Livestock Life Cycles

Note: All numbers are averages only.

	CATTLE	SHEEP	GOATS
Gestation	285 days	150 days	150 days
Castration	Birth to 4 months	By 3 weeks old	By 3 weeks old
Vaccination	Calves: just before weaning. Adults: annually	Lambs: by 8 weeks. Adults: annually 2–4 weeks before giving birth	Kids: by 8 weeks. Adults: annually 2–4 weeks before giving birth
Weaning	Variable; most beef calves at 4–6 months, most dairy calves at birth	2–6 months	3–6 months
Sexual maturity	9–16 months	5–12 months	2–8 months
Age at first breeding	At 70% of adult weight; 13 months at earliest	At 70% of adult weight	At 70% of adult weight
Cycle length	21 days average, year-round	17 days, in fall	21 days, in fall
Butchering	Most at 15–24 months	Less than 9 months	Less than 9 months

CHAPTER FOUR

Mid-Spring

Soil temperatures reach 50°F/10°C and continue to warm through the season. Cool-season grass growth accelerates; field corn is planted.

THE SOIL IS WARMING, the grass is taking off, the chicks are feathering out, there are calves and lambs on the ground, and it's time to finish planting fields and most of the garden. This is the season, with its mild weather and clean grass, that I like to have my cows calve (which means I had to turn the bull out last July). It's time for rhubarb sauce, made from the first fruit of the year.

Keep an even closer eye than usual on the weather so that you time planting and weeding or cultivating around rainfall — if you can. It's good to plant any seed that is broadcast, such as when overseeding the wintering area, just before a rain, or even during a light rain. The rain will press the seed into better soil contact, improving germination. It's best to do hand-weeding after a rain, when roots pull up easily, but wait until the soil is a little dryer to use a mechanical cultivator.

Mid-Spring

FOCUS ON:
garden • barn

Garden

» **Plant seeds in the ground** according to their temperature preferences. It's primary planting season.

» **Harvest rhubarb.**

Field

» **Plant row crops,** such as field corn.

Pasture

» **Move livestock** through large paddocks as forage growth accelerates, letting them just take off the tops to keep grasses and clovers from maturing too quickly.

Orchard

» **Start monitoring for pest and disease problems now** if you intend to actively control them. Observe trees, measure degree days, and place pest-specific traps.

Beeyard

» **Feed new hives as needed** until well established. The biggest nectar flow of the year is getting under way as fruit trees and many other plants blossom.

» **Observe hives regularly,** but leave them alone unless a problem is evident.

Barn

» **Calving, lambing, and kidding** start or continue.

» **Buy and bring home weaner pigs.**

» **Castrate and vaccinate** lambs and kids.

» **Castrate calves** now or in fall.

» **Arrange to have sheep sheared** by early summer.

Coop

» **Place chicks outside** as soon as they're covered in feathers.

» **Check for eggs daily.** Egg production in the older hens accelerates when they're out each day to bask in the sun and feast on bugs.

Equipment Shed

» **Clean up and store planting equipment** as you finish using it.

» **Pull out cultivating equipment.**

Woodlot

» **Monitor for invasive forbs** and pull them before they go to seed.

» **It's prime nesting season for birds.**

Wildlife Habitat

» **Plant native flowers** for butterflies and beneficial insects.

» **Identify and record what native plants** are already growing on your land.

» **Plant food plots for wildlife.**

Garden

If you planted the most cold-tolerant vegetables as soon as the ground could be worked, then the first greens and radishes will be on the table this season even as you continue to plant the rest of the garden. Start with the more cold-tolerant seeds and progress as the soil warms to those that need more congenial temperatures. Include trap crops to distract harmful insects and flowers to attract beneficial insects (for a discussion of the purpose and process of developing habitat for beneficial organisms, see pages 83 and 141). If it's dry, you can mulch, but if it's wet you might want to wait — wet mulch on wet ground is great habitat for mold and slugs.

Field

If the weather has cooperated, you should have finished spring tillage and planted small grains by this time. The only thing left to plant is row crops, which need warmer soil than most of the small grains and forage crops.

As soon as you've cleaned and stored planting equipment, bring out the cultivator (mechanical weeder). In an organic system, cultivating typically begins as soon as the crop is up, though you may have already been over the field a couple of times before planting. Continue cultivating until the leaves of the crop plant meet between the rows, shading the soil enough that weeds won't succeed (also called canopy closure). Another method is to plant a living mulch between the rows to smother weeds, feed the soil, and minimize mud when harvesting equipment moves through. The timing of planting a living mulch depends on what row crop you've planted and how fast it grows, and what mulch you plan

Mulch

Mulch conserves moisture, keeps the soil temperature more even, suppresses weeds, and keeps people and equipment out of the mud. When organic mulches rot, they can be tilled into the soil to add organic matter.

When you apply mulch to garden paths first thing in spring after tilling, the rest of the spring work is much cleaner. Wait until the weather starts to become dry to mulch between rows and plants, so that you minimize shelter for the slugs, bugs, and diseases that thrive in a cool, wet environment. Soaker hoses or drip irrigation lines run under the mulch help to keep the top of it dry through summer.

Our favorite mulch is hay — old or new — or straw from the small grains. The NOP Final Rule allows the use of plastic mulch (sheet plastic) as long as it's removed at the end of the season.

An antique corn planter may be just the right size and price for a small farm, if you can get it running.

to plant and how fast and high it will grow — you don't want it competing with the main crop for sun and moisture. Some examples of how to figure out the best timing are given on page 82. Organic mulches can be applied on small fields most easily with a mechanical blower (if you can rent or borrow one) that will blow it between the rows.

Pasture

During mid-spring, grass growth generally goes from dawdling to explosive, and livestock should be moved in sync with the growth — slowly through big paddocks at first, then more quickly through smaller paddocks to "top off" the grass to ensure it doesn't get overmature and less palatable. As grass growth slows later in summer, slow the rotation as well, so the livestock eat more of the grass in each paddock before being moved (but don't let them graze it below 4 inches if you want a timely recovery). Walk pastures regularly from now until the end of the growing season to judge the rate of growth and quality of forage (mature seed heads indicate it's too old for good eating), then move animals as dictated by grass growth and their age or class. Some tips for moving livestock:

- If the water trough is in the barnyard and the herd or flock comes up to it at a regular time each day, change the gates while the animals are in the barnyard and let them move themselves to the new paddock.

- Do not move livestock first thing in the morning; they'll start expecting you to be out there at dawn (which comes very, very early in midsummer), and if you don't show up they will stand at the barnyard gate and make a lot of noise. Move them in the afternoon or after morning milking.

- Carry a white 5-gallon bucket with a little grain or treats from the garden or orchard whenever you need the livestock to come or follow. Dole out the treats as you lead them. They'll learn to come at the sight of the bucket.

- Call to the livestock when you go out to move them to fresh grass; they'll learn to come at the sound of your voice.

Orchard

Many fruit trees, especially apples, begin breaking into blossom at this time, setting in motion the year's biggest nectar flow. But bees aren't the only insects you'll see — by now bug season is well under way. Unless you're having major insect and disease problems in your orchard every year, all you need to do is keep tent caterpillars under control. If trees are on a good site, well spaced and pruned, with plenty of habitat for beneficial birds and insects, there will be enough quality fruit for eating and the poorer stuff can go into jellies, fruit butters, and juice.

If you prefer to be more proactive about preventing insects and disease in your orchard, use IPM (integrated pest management). This involves learning to identify pests and diseases and their life cycles and being proactive to prevent them from becoming a problem. Regularly scout your orchard to identify potential problems, and set up monitoring traps and other traps before it's time for an insect hatch or a disease flare-up. For specifics on pest and disease control in an orchard, see page 116; for more on the general principles of IPM, see page 141.

Beeyard

With blooms busting out all over, it's best to observe the hives daily as you make your rounds, to make sure the bees are busy during the biggest nectar flow of the year. Open the hives (preferably around noon on a sunny day when the bees are busy and happy) to see if a super should be added. The rule of thumb is to add the first one when 7 of the 10 frames in the hive body are full of eggs and brood. Add additional supers early and often rather than late, to discourage swarming caused by overcrowding. New hives may still need to be fed at this time; keep syrup available until the bees quit paying attention to it.

Bee activity increases and bee numbers build rapidly at this time of year.

Barn

Though livestock are on pasture now, keep hay available until they decide they don't need it anymore. If you're feeding a grain ration, they may even lose interest in that for a week or two at this season. Unless they're showing other signs of illness, don't worry — it's normal for them to binge on lush, young grass at this time. Do monitor the consistency of their manure; if it looks like cake batter, that's okay, but if it's liquid, they either have a digestive upset or will soon if you don't add fiber to their diet. This is especially important with the youngsters: diarrhea, no matter what the cause, can kill them quickly.

Once settled into their sty, pigs should be given daily access to the outside as well. You can do this by fencing in an area adjacent to a permanent sty and subdividing it so the pigs can move to fresh ground every week or so. Or you can build a portable pig hut, put them farther afield, and move them more often. Young pigs will clean up the June drop apples and save you the job.

Pigs will attempt to both dig under and crawl over fences they can't push their way through. Nonelectric pig fences should be sturdy and at least 4 feet high, with barbed wire at ground level to discourage rooting. Electric fences are simpler and very effective, if regularly checked to make sure the pigs haven't rooted a pile of soil over the wire.

Planning for Winter

With youngsters on the ground and breeding seasons approaching, this is a good time to observe livestock with an eye toward deciding which animals will be sold, butchered, or bred in the months ahead. Consider butchering or selling at auction any animals that did not have

Basic Pest and Disease Control for Animals

Flies and parasites are at their most aggressive during the grazing months, and disease knows no season. It's important to be proactive about preventing serious problems. Start with these basic steps:

1. **Provide a healthy environment.** Provide clean water and bedding, adequate and appropriate feed for the species, salt and mineral mix appropriate for the species, shelter from the elements, and free access to green pasture.
2. **Practice good sanitation.** Minimize the mud, manure, old wet hay, soiled bedding, and puddles where insects breed. Clean indoor housing for pigs and other penned animals daily, and clean chicken coops at least twice a year. Keep weeds trimmed along buildings and fences to eliminate habitat for flies. Run chickens with or behind livestock to clean up insects and their eggs.
3. **Vaccinate.** This is a cheap and simple way of preventing several diseases that can cripple or kill your animals. If you choose not to vaccinate, you may not legally be able to show your animal or transport it across state lines.

young or that have bad temperaments. If you plan to direct-market meat animals, ask your customers now for their fall orders.

If you plan to breed your livestock, start visiting seed stock producers to select a bull, ram, or billy. Arrange for delivery at the appropriate time. The sooner you get going on this job, the better selection you'll have.

Coop

As soon as the chicks have a nearly full covering of feathers, they can go outside, as long as the weather isn't unseasonably cold. I wait until nighttime temperatures are mostly in the 50s (10 to 15°C) and daytime temps are at least in the 60s (16 to 20°C). A couple of cold days or nights won't do any harm, but the birds can become miserable and quit growing if they're outside during a prolonged cold spell. Young chickens just don't have enough body mass or thick enough feathers to tolerate cold, especially if it's windy.

Meat chickens' lives are too short to get into the habit of roosting at night, so rather than giving them a regular coop, it's best to raise them on pasture in bottomless pens. These clean and humane pens are an old technique made popular again by sustainable farmer and author Joel Salatin, with the direct result that there are a lot more happy chickens around than there used to be. The pens are traditionally 10 by 8 feet and about 2 feet tall. The front half of the top is hinged so you can put in feed, grit, and water and catch chickens when you need to. The top and sides of the back half are covered with lightweight tin or plastic for shelter in bad weather. The pens are moved daily to a new patch of grass. A week or so after a pen has been moved, the grass that was beneath it becomes green and vibrant again due to the application of chicken manure. I used to move our pens by putting skateboards under the back, then lifting the front slightly to roll it forward. An Internet search for "pastured poultry pens" will yield many alternative pen designs.

Laying hens are different. They'll be around for at least two or three years and need a secure home in summer and winter. When they're ready to move out of the brooder, put them in the coop and keep them inside for three days. Then, to train the young hens to come home at night, let them out for just an hour at first, right

Portable poultry pens come in many sizes and shapes. This is a classic Salatin pen pioneered by sustainable farming guru Joel Salatin.

before sunset, so they don't get too far away and get lost before bedtime. Gradually increase this time day by day until they're accustomed to being out all day and coming home at night.

During summer, laying hens can be kept on pasture in a portable coop surrounded by portable electric poultry netting to keep predators out. Move the coop along an electric fence line, so you can attach the poultry fencing to that without having to run extra wires or an extra energizer. For just the few hens it takes to keep the family supplied with eggs, I've found I take better care of them if they're kept in the permanent coop and allowed to run free in the yard. And as a bonus, they keep the yard tick-free.

Equipment Shed

As you finish planting, clean up and store grain drills and planters, because storing equipment immediately after use will greatly extend its life. It's much easier to clean up dirt or dreck when it's fresh. That said, don't use the power washer on planting equipment — any water that settles into the tubs can collect dirt or cause rust, and that keeps the seed from sowing smoothly. Instead, blow off these implements with the shop vacuum or an air compressor. I've also learned to pull the haying equipment out before I put away tillage and planting equipment, since that's what I'll need next and I don't want to have to move everything again to get at the mower. But never leave a small square baler outside and uncovered for any length of time — rain can play havoc with the twine-knotting mechanism.

We, like many farmers, keep an area between the house and fields mowed short so we can stage equipment there. We'll pull out the equipment and hook it up a few days before we'll need it, and use that grass strip as a place to park stuff in the evening while it's still in use. This is much easier than pulling things in and out of the shed every day, and the light's much better for checking things over.

Woodlot

Unless the flies, mosquitoes, and heat make woods work intolerable, continue to pull or kill invasive species, especially the nonwoody invasives that have emerged in the spring weather, before they can set seed and multiply your problems. Make sure the plants are dry and dead before composting them, or they will come back to life. You can also burn them.

Wildlife Habitat

Native plants that are leafing out and blossoming may be easier to identify now than at any other time of year. Jot down what you've found to help decide what more to add. Plant native flowers for pollinators and beneficial insects along field edges, in the orchard, and in the garden. Check any grass and forb seedings you made earlier to see if plants are up. Weed if needed.

Plant food plots for game animals at this time, possibly in an opening in the woods. Till up the ground as for a garden, though the ground might be quite a bit rougher due to roots left after clearing brush and trees last fall. Buy seed mixes at farm stores and seed dealers.

Topic of the Season

Rotational Grazing

Rotational grazing, more correctly but less commonly called management-intensive grazing, is a proven method for improving pastures, growing more forage through more of the year, and lowering the cost of raising livestock. The technique mimics the natural behavior of wild grazing animals, which have evolved a mutually beneficial relationship with the forage plants they depend on. An additional and very significant benefit is soil improvement from the even manuring and the increased abundance of roots that hold soil and add organic matter. Livestock, correctly applied, are a fantastic tool for building soil.

The general idea of rotational grazing is to put a large number of animals in a small area for a short period of time, then remove them for a long period of time. This ensures that plants are evenly grazed, fertilizing manure is applied fairly evenly on the soil, and the grasses and clovers have plenty of time to regrow both tops and roots. The result is a much more constant and plentiful supply of palatable, nutritious green feed throughout the grazing season.

To divide the traditional large pasture into smaller units, it's cheapest and simplest to use temporary electric fencing. Temporary electric fence is remarkably quick to set up and take down, and it gives you maximum flexibility when varying paddock sizes and configurations.

The length of the rest period for paddocks should be appropriate to your climate and the season. Pastures in regions with relatively cool, moist summers need as little as 10 days' rest at maximum grass growth in spring, and 3 to 4 weeks during the "summer slump" when the weather is at its hottest and driest of the year. In dry regions like much of the West, rest periods are many weeks longer. When judging how often to move animals on your acres, look at the grass growth rate and the amount of rain you're getting, not the calendar. If it has gotten so dry that grass is not growing at all, pull the animals off pasture and feed them hay.

In addition to allowing you to raise more animals on less land with better results for both soil and livestock, rotational grazing can create savings on the annual feed bill. By moving animals swiftly through paddocks in early spring when the grass is just beginning to grow and reserving some paddocks for additional grazing after the growing season is over, the grazing season can be extended at both ends, and you'll need less hay.

How Grass Grows

Unlike trees and many other plants, a blade of grass grows from the bottom, not the tip. As the grass gets higher, it shades the growing point near the soil surface, slowing and eventually stopping growth. If it's not cut, the sod thins and browns. Graze the grass to put sunshine on the growing point, and it regrows quickly from its big root system, but graze the grass too often and the root system becomes exhausted.

Wild herds have evolved to graze in a bunch in order to be less vulnerable to predators. They will graze an area closely, then move on and be gone from that patch for long periods, giving the grass time to recover. Rotational grazing, where livestock are rotated through a series of paddocks, mimics this system. When managed correctly, rotational grazing results in thick, lush pastures through the growing season.

Cool-Season vs. Warm-Season Grasses

Grasses fall into two general categories: cool-season grasses that grow best between 65 and 75°F (18 and 24°C), and warm-season grasses that take off when temperatures get into the 90s (32 to 37°C). In the northern states, there's typically a "summer slump" when the cool-season grasses characteristic of those pastures find it too hot to do much growing. In the South, there's a "winter slump" when the warm-season grasses are chilled into inactivity. If the slump is long enough in your locale, you can plant separate pastures with the other type of grass to fill the gap. Warm-season grasses especially need adequate rest between grazings to thrive, as well as enough time to regrow to a height of 6 to 8 inches before winter. Get professional or neighborly advice on what varieties do best in your area and on your soil type, and about when to plant, graze, and/or cut for hay.

Warm-season grasses good for pasture: Big bluestem, switchgrass, Indian grass, other natives

Cool-season grasses good for pasture: Brome, fescue, orchard grass (bluegrass is good too, but doesn't produce nearly as much tonnage)

Laying Out a Grazing System

When subdividing your pasture land into smaller paddocks for a rotational grazing system, keep these principles in mind:

1. Paddocks should enclose (where possible) a uniform piece of land — a hillside, a flat area, or land that all faces the same direction — for the most even grazing and recovery. Try to put fences on break points between flat and sloped land, or between slopes that face different directions.
2. Build enough paddocks to ensure that the vegetation in each one has an adequate recovery period before animals return to it. For example, if I plan for an average rest period through the season of about 21 days, and grazing periods of 3 days per paddock, I'll want 8 paddocks — 21 days divided by 3, plus the paddock that they're using, which isn't resting.
3. Paddocks should generally be sized to accommodate your animals for no less than 12 hours and no more than several days.
4. Build lanes to move livestock past closed paddocks to far ones. The gate from a lane into a paddock should be in the corner of the paddock closest to the barn and/or water supply, since this is the direction the animals will naturally want to move.

To accommodate the changing rate of grass growth through the season, you can:

- Change the paddock size, thereby reducing or increasing the overall number of paddocks and promoting more or less intensive grazing to ensure adequate rest periods.
- Subtract paddocks from the rotation during rapid grass growth and make hay instead, or add paddocks during slow periods and make less hay. (I start with 8 paddocks; at the end of summer, I typically have 12 to 17, with the extra ones created by adding in some hay ground.)
- Feed hay in dry periods rather than overgraze paddocks to the point where the soil is exposed and vulnerable.

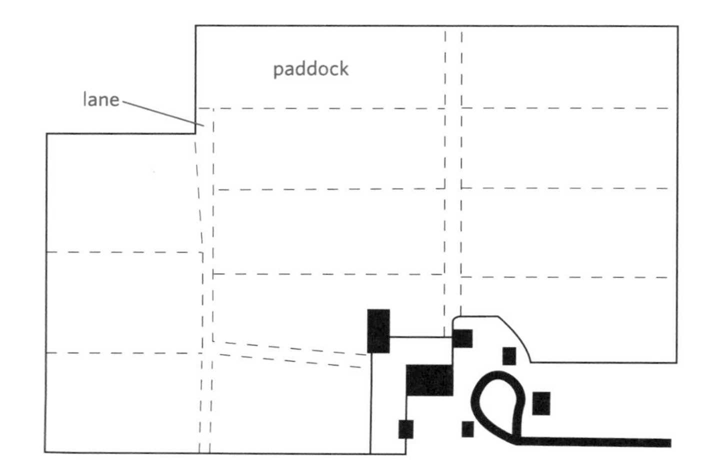

This sample grazing system shows lanes and paddocks.

How Tall Should the Grass Grow?

After following the topic closely for 20 years or so, I've concluded that there are as many ways of managing grazing as there are people doing it. Here are some thoughts for helping you figure out the details of your grazing system.

Lush, young grass produces the most meat and muscle quickly, but the lusher and younger it is, the more prone your livestock will be to digestive problems. It's like eating nothing but steak and chocolate all the time. The exceptions are poultry (young grass is excellent for them) and, in general, young stock being grown or fattened, but supplement with hay if their manure is very runny.

Taller, older grass produces more forage per acre, puts more fiber in the diet, and is good for digestion and overall health, though it doesn't grow as much meat or milk. It is excellent for equines, milk animals, and beef cattle, and fine for all livestock — though you may want to supplement with grain for better meat and milk production.

Mature grass with dry seed heads is not very palatable, and most livestock will reduce their intake rather than eat too much of this. At this point it's best to mow the paddock to renew the grass.

You should restrict your equines' grazing to prevent them from becoming too fat. They are also prone to colic and founder when turned out suddenly on lush grass, or if their overall diet is too rich, with grass that is too young to have enough fiber. Instead of moving them to fresh paddocks, use strip grazing — move a fence a few feet each day to give them a regular amount of fresh grazing.

Using Temporary Electric Fence

Electric fencing, either temporary or permanent, is effective for managed grazing and protecting orchards and gardens from four-footed wildlife if correctly configured. The essential components of a temporary electric fence system are a fence energizer, lightweight fence posts that can be pushed into the ground by hand or foot, and lightweight plastic wire or netting containing metal strands that can be quickly reeled or unreeled to create temporary paddocks.

It's best to have a permanent fence of wood or wire around the perimeter of your pastures, since electricity, though very effective, is not reliable if a storm knocks out the power, a branch falls on the line, or a stray dog decides to chase your livestock. Running an electric wire along the inside of the perimeter fence gives you maximum flexibility when you're subdividing the pasture with temporary fence — you can hook it up anywhere along the perimeter.

Setting up and taking down temporary electric fence is quick and simple: walk or drive your ATV down the line, pushing in posts as you go, then on the way back unreel the wire and snap it into the clips on the posts. Reverse the process to take it down. I recommend using metal T-posts at corners that aren't formed by a connection to the perimeter fence; a step-in post won't stand up against the pressure of the wire pulling in two directions. Or you can install three step-ins on the corners.

Your electric fencing is only useful if it is properly constructed and powered. Diagrams

Gates in electric fences can be simply made with purchased gate handles that hook into a loop of fence wire.

Mid-Spring

for how to connect the entire system are easily available from fencing supply companies, but here are a few additional tips:

- The energizer should, as a rule, be oversized for the length of fence wire you'll be running, since vegetation on the wire and all the cutting and knotting you end up doing increase the electrical resistance.
- Use made-for-the-purpose run-out wire to connect the energizer to the fence; anything else can melt.
- Use more and longer ground rods than recommended to ensure an adequate circuit when the ground is dry or frozen.
- Don't depend on hoof-to-ground contact for completing the electrical circuit when the ground is dry, frozen, or snow covered: it won't work. Instead, install a grounded wire between the live wires on the fence, close enough so that if an animal sticks its head through the fence, the circuit from live to ground will complete itself across the animal's hide. (*Note:* If you use electrical netting, get the kind that has built-in ground wires.)

Fitting the Livestock to the Land

All livestock and poultry should have access to the outdoors and green grass to live happy, healthy, productive lives, but different species need very different amounts of land. Chickens and pigs can be raised in quite small areas, while a cow and calf may need anywhere from 2 to 40 acres to have enough pasture and hay to get them through the year. A common rule of thumb is that five sheep or goats can live on what it takes to support one cow.

You'll probably also want to consider the initial investment and the amount of labor involved. Chickens are the cheapest to obtain, can be housed in coops built from scrap wood, and don't cost too much to feed, especially if you're letting them run outside and giving them table scraps. Weaner pigs are relatively cheap. Pigs need more room than chickens but not much, though they require good shelter from both heat and cold and very good fences. On the other hand, they go from piglet to freezer in 6 months and you have the winter off, unless you decide to go into the pig production business for yourself.

Sheep and goats cost more, require good fences and adequate shelter (especially goats, who don't tolerate cold well), but they are small enough for easy handling. They're much easier to move around than pigs. Cattle cost the most, need the most land, and take the longest to mature, but are the easiest to fence and the least likely to be bothered by predators. They're the most weather-tolerant of livestock, though they need to be able to get out of the worst heat and coldest winds. Unless they're halter-broken, you'll want some well-built handling facilities — at the minimum a good headgate.

CHAPTER FIVE

Late Spring

Soils reach a temperature of 60°F/16°C and warm through the season. Cool-season grass growth explodes; soybeans, sunflowers, and sorghum are planted. First cutting of alfalfa gets under way.

BY LATE SPRING, both air and soil temperatures are reliably warm and insect populations are exploding, but you can walk outside without a jacket on a sunny morning with a light breeze and the bugs aren't a bother. Steamed fresh asparagus and rhubarb sauce grace the dinner table, and grazing, gardening, and mowing are all moving along in high gear. By late spring you're focused on keeping everything growing and healthy. The planning is over for the moment; now it's all about hitting the ground running in the morning and getting the work done.

Late Spring

FOCUS ON:
garden • pastures

Garden

» **Plant the most tender seeds** and begin succession plantings as the soil warms to 70°F/21°C and higher.

» **Control weeds early** and often.

» **Pick early vegetables.**

» **Transplant most indoor seedlings** to the garden, but have covers ready for the tomatoes, peppers, and squash in case of frost.

Field

» **Finish planting.**

» **Start cultivating row crops** when they're a few inches high.

Pasture

» **Finish putting up temporary fence** and keep the animals moving.

» **Control weeds** by mowing.

» **Set aside some paddocks** for making hay and speed up the rotation if the grass starts growing faster than the livestock can keep it grazed. Grass growth is explosive.

Orchard

- **Thin apple blossoms** if desired.
- **Continue to monitor for insects** and disease.

Beeyard

- **Add supers** as needed.

Barn

- **Put all of the livestock** out on pasture.
- **Calving, kidding, and lambing** finish.
- **Find a bull** or an artificial insemination service for breeding the cows.

Coop

- **Place meat chickens** in pens out on pasture.
- **Let out laying hens** during the day.

Equipment Shed

» **Clean and store** planting equipment.

» **Tune up cultivators.**

» **Pull out** haying equipment.

Woodlot

» **Control invasive species.**

Wildlife Habitat

» **Continue observing growth** of resident native species and control weeds in plantings.

» **Avoid making hay** or mowing pastures where ground-nesting birds are present.

Garden

After planting steadily since early spring, the first round is finishing up at this season with the soil warm enough to seed pumpkin, squash, and watermelon. The air is warm enough to move the last of the inside seedlings to the garden: peppers, tomatoes, and eggplant. But all danger of frost is not past, so keep buckets or blankets handy to cover those plants on nights the temperature might dip below 32°F/0°C. In areas with long growing seasons and patient gardeners, it might be better to keep those seedlings indoors until all danger of frost is past.

To ensure a continuous supply of young greens, peas, and other crops through summer, you can plant more of these now, and again every couple of weeks until late summer. Vegetables that dislike hot weather can be harvested before the worst of summer heat, and more can be planted in late summer for a fall harvest — this works especially nicely with cool-season greens and the faster-growing root crops.

Weeds will be growing as fast as the vegetables, and hoeing or cultivating is a continuous job at this season, both between and in the rows or beds. You can apply a light mulch at this time if the weather is getting drier, or plant a cover crop between the rows once the vegetables are tall enough not to be overtaken.

Hardening Off

Don't transplant seedlings into the garden before taking a few days to get them used to full sun and wind. Start by moving them outside for 1 hour the first day, then 2 hours the second, 4 hours the third, 6 to 8 hours the fourth, and from sunup to sundown on the fifth day (if the weather cooperates). They should be ready to go into the garden the following day.

Mulch

Mulch is great: on garden paths it keeps you out of the mud; in between and around plants it suppresses weeds, conserves water (a lot!), and keeps ground-level fruits and vegetables clean. But when the weather is wet, mulch can also encourage slugs and mold. Hold off on mulching until the weather turns dry. If the weather remains wet throughout the summer, use just a light mulch (meaning you can see soil) to still gain some benefits without encouraging the mold and slugs.

Till mulch in at the end of the season before planting winter cover crops, or leave it to protect the soil and till it under in spring. If it's too bulky and tough for tillage, rake it off and compost it.

Field

Once you've planted the row crops, focus on cultivating. Good weed control is important not only for this year's crop but for next year's as well, especially if it will be a small grain or forage crop. How often you'll cultivate is dependent on how much weed pressure there is and how much time you have, but it's generally better to remove the weeds when they're younger rather than older. Take your time and pay attention; you don't want to accidentally take out a few rows at the corner and create a case of "cultivator blight."

Cultivating Implements

There are three basic types of cultivators: those that cut off the roots from the tops of weeds, those that uproot weeds, and those that bury them. Some implements are best for between the rows, while others can be effective for in-row weeds, depending on the crop. Implements that bury in-row weeds bury the salad greens, too, for example, but they work well with corn and potatoes, since both of those crops do better when you hill them with soil.

You can also weed by hand, using your fingers or any one of the many varieties of hoes available, which operate in the same ways as mechanical cultivators.

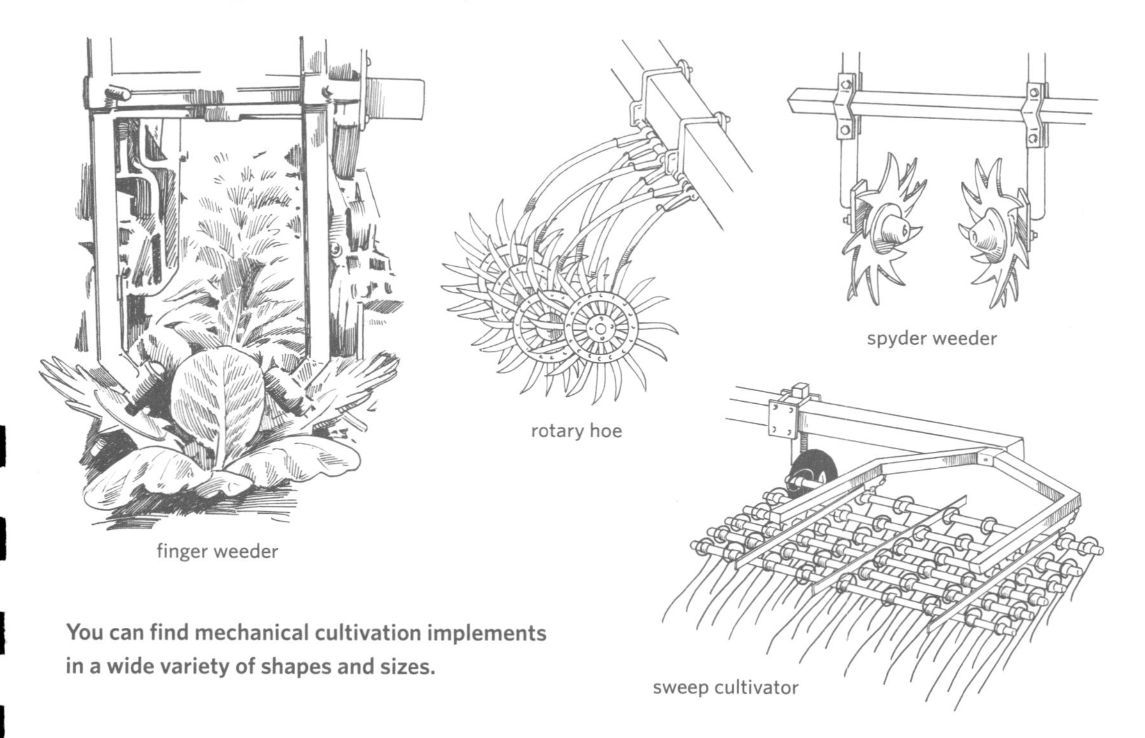

You can find mechanical cultivation implements in a wide variety of shapes and sizes.

Livestock Parasites

Livestock parasites, both internal and external, have evolved life cycles that take advantage of the natural behavior of grazing livestock. The frequency of paddock shifts and the length of rest periods in a rotational grazing system can exacerbate parasite problems when they synchronize with parasite life cycles. If your herd or flock is thin with dull coats, and manure tests by a veterinarian indicate a heavy parasite load, you must take action. The first step is to change the length of the pasture rotation by at least 1 week in either direction, so that animals are not returning to a paddock when the new generation of parasites is at its peak.

The most prevalent internal parasites in grazing animals are several species of intestinal roundworms (found throughout the United States), tapeworms, liver flukes, and lungworms. Take these steps for control:

1. **Identify the problem.** If animals are thin, rough-coated, and slow to gain weight, suspect parasites. Take a manure sample to the vet to have it examined for worm eggs.
2. **Learn the life cycle of the target species.** In general, parasite eggs are shed with the manure. They hatch if it's not too cold or too hot and not too dry. After a predictable number of days, they reach the larval stage, when they crawl onto grass blades and are ingested by grazing animals, and the cycle begins again.
3. **Choose one or several options to prevent reinfection:**
   - Change the length of the pasture rotation so that livestock don't return to a paddock when the parasite is on grass blades waiting to be ingested.
   - Don't force livestock to graze too closely to manure pats — move the animals to fresh pasture before they get that hungry.
   - Follow one type of livestock with another, since parasites are specific to a species and will die if ingested by the wrong kind of animal.
   - Run chickens through the pasture with or after livestock to break up manure pats and eat the eggs of parasites or expose them to the sun.
   - If you use a dewormer, treat twice in spring 3 weeks apart just after turnout. In the South, do the same in late summer or early fall along with a single deworming in spring. Alternate deworming products to slow the buildup of resistance in worm populations.
4. **Lastly, monitor the results.** Test manure at least annually if you suspect parasite problems.

Pasture

As the rate of grass growth peaks for the year and youngsters are just getting used to electric fences, keep the paddocks big and don't graze them too hard — at this time, it's more important to keep the grass from becoming overmature than to utilize all of it. If the grass gets ahead of the animals, pull some paddocks out of the rotation and make hay, or at least mow them.

Weed Control in Pastures

By now weeds (defined for this discussion as plants that your livestock won't eat) are becoming obvious out in the pasture, and ideally they should be controlled before they set seed and multiply the problem. In an organic system, the best options are mowing or hand chopping. Mowing the weed patches, or the entire paddock, will freshen up the paddock nicely, and at this time of year you generally have grass to spare. On rough pastures where you can't move a mower, you can use a handheld chopper (it looks like an overgrown golf club) to keep things under control. Mowing and chopping don't take care of the roots, so the weeds will come back.

Once the garden is in and before haymaking, I also like to brush mow the fence lines. It is especially important to do this before the coarse grasses that grow in patches there get too rank and fibrous for the mower to handle easily.

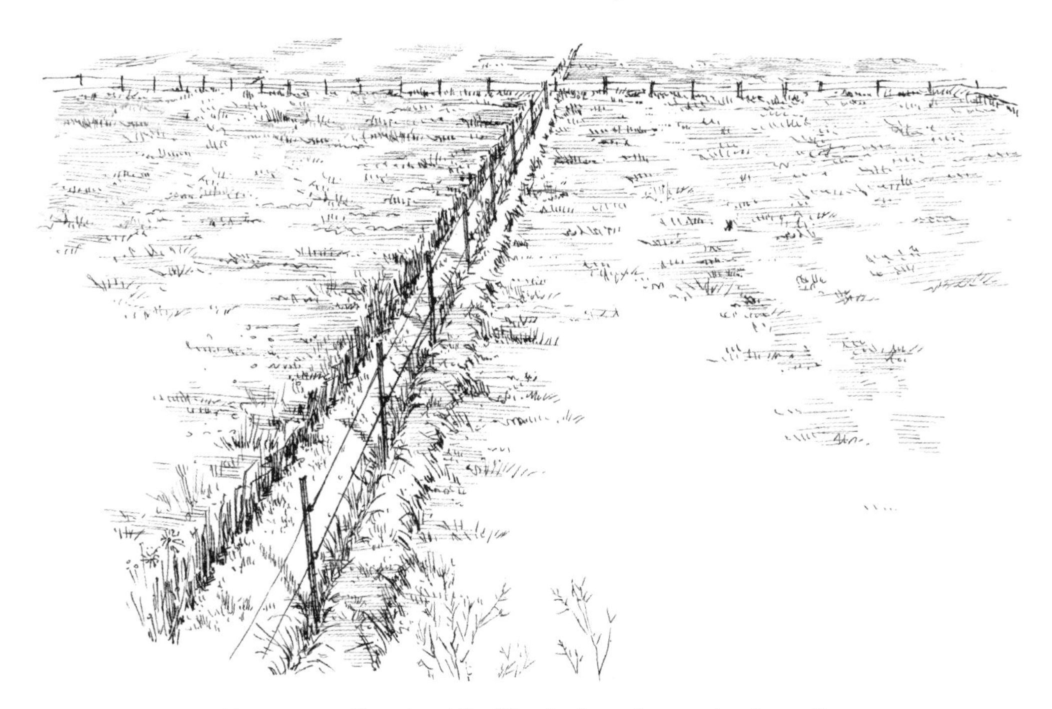

You can greatly extend the life of a fence by mowing fence lines to keep brush and vines off the poles and wires.

Orchard

As apple blossoming time peaks, enjoy the show while you continue to search out and destroy tent caterpillar nests. If you want to thin the apples to even out the harvest from year to year (many varieties of apple trees tend to bear a good crop only every other year), this is your first opportunity to do so. Hand thin the blossoms if you have just a few trees, or chemically thin them. Organic orchardist Jim Koan of Michigan uses salt water or lime sulfur, carefully applied.

Beeyard

If you medicated your bees and added supers in mid-spring, there's little to do at this season except keep an eye on the hives for any signs of problems, such as too little activity or overcrowding, which can lead to swarming. New hives should be fed until they quit coming to the feeder. If you didn't add supers because the hive body frames weren't full, then add them now.

Barn

Calving, kidding, and lambing should be finished by now (though one or two youngsters often arrive late), and aside from daily checks to make sure all are present and healthy, the livestock shouldn't require too much attention at this time of year. Keep salt and mineral feeders freshened, and when the water tanks start turning green, dump them out and scrub them. Our old farming neighbors tell me they used to keep a couple of carp, caught during their spring run, in the stock tank all summer to keep it clean.

Find a bull or an artificial insemination service for breeding your cows in midsummer. A bull is the simplest way to get the job done if you have more than a few cows; just make sure he's fertility tested before delivery. If you have just a few cows, and they're accustomed to being tied or being in a chute and headgate, it might make more sense to artificially inseminate and skip having to deal with a large, expensive, and not entirely trustworthy animal.

Babies and Fences

If your fences are horizontal wires instead of mesh, quite often a calf or lamb will lie down next to the fence and get up on the wrong side of the wire. You'll find out about it when you hear the noise made by mother and baby who can't figure out why they've been suddenly separated or how to fix it.

To avoid spooking the calf and having it run away, I've learned to wait a couple of hours, until mom's udder or bag is uncomfortably full and the baby is starting to think that if it doesn't get to eat soon it will starve to death. At this point, I go out, open the nearest gate, get on the side where the lamb or calf is, and (not too close or too fast) drift it along the fence line to the gate where, if all goes as planned, the two will be happily reunited on the right side of the fence. I have learned to even let a calf bawl all night (they are surprisingly tough animals) rather than try to move it in the dark and have it panic and run off to who knows where.

Coop

As with other livestock, poultry need little attention at this time of year besides checking them daily and keeping grain feeders, grit feeders, and waterers clean and full. Move the pastured meat chickens to new grass once a day, and let the laying hens out in the morning and shut them in at night. If your hens start laying everywhere but inside the nest boxes in their coop, then keep them in until noon or so to encourage them to lay before you let them out.

Equipment Shed

As the hay crop begins to mature, pull out your equipment and make sure it's all clean and running — a winter in the shed is often an invitation for mice and pigeons to nest in the implements. Keep the blades on the lawnmower and brush mower sharp.

Woodlot

With nesting season and bug populations at their height, and plenty to do elsewhere, we stay out of the woods in late spring unless we're pulling invasive species.

Wildlife Habitat

Continue to observe, identify, and record native plants and animals on your land. Check birdhouses to determine what species are nesting there and whether eggs or nestlings have been destroyed by predators. Avoid mowing hayfields and pastures where ground-nesting birds are present; wait until you can see the fledglings perched on fence wires.

Build birdhouses to the specifications for the species you are trying attract. Otherwise, the houses may become inhabited by less desirable birds.

Cover Crops

Cover crops are any crops that are not planted for the primary purpose of human or animal feed. Cover crops are a critical component of the crop rotation, especially for field production on a sustainable organic farm, since they boost soil health and fertility and serve other purposes that on conventional farms are addressed with chemicals or generally ignored.

Cover crops are often loosely subclassified by their primary purpose: green manures, living mulches, winter cover crops, and trap crops and beneficial habitat plants (to distract insect pests and attract beneficial insects and birds). In reality, all cover crops will improve the soil, prevent erosion, and help to control weeds, insects, and soilborne disease to a greater or lesser extent.

When to Plant Cover Crops

Cover crops can be planted:

- In early spring to be tilled in as a green manure before food crops are planted
- In early spring to serve as a nurse crop for a forage legume
- In spring to occupy bare ground throughout the growing season
- After main crops are up to serve as a living mulch between the rows
- In early to midsummer to fill the gap between early harvested vegetables and vegetables planted for fall harvest
- In late summer to serve as a nurse crop for a forage legume
- After small grain harvest to serve as a green manure and/or winter cover
- Several weeks before or after row crop harvest to serve as a green manure and/or winter cover, to be tilled under or planted into in spring

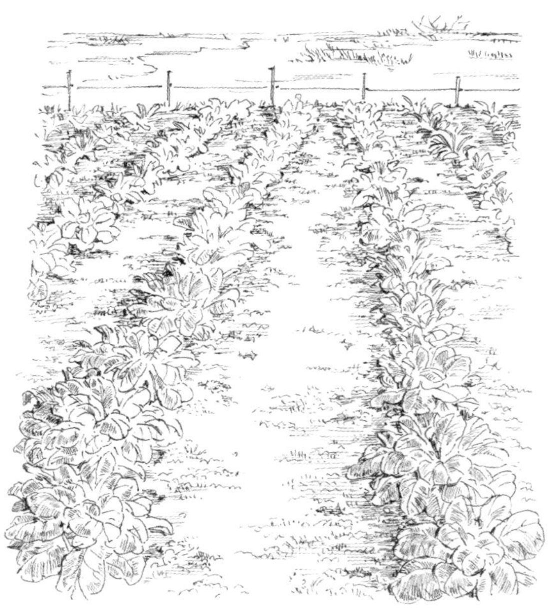

Plant white clover between rows of cabbage to minimize mud and dust and add nitrogen and organic matter to the soil.

The plants most commonly used as cover crops are small grains, grasses, and legumes. Winter grains and hardy legumes will not die over winter; when planted in fall, they are generally plowed down as a green manure in spring before the food crops are planted. Spring grains and annual legumes planted in fall will die over winter, but they still serve as an erosion-preventing mulch. Like vegetables and field crops, cover crops vary in their heat and cold tolerance and their adaptability to different soil types and moisture levels. Some cover crops can also be used as season-extending pasture for livestock or poultry.

You can find an excellent summary of how and when to use different cover crops for different purposes in different regions in *Managing Cover Crops Profitably*, available from the Sustainable Agriculture Research and Education (SARE) program supported by the USDA. Below is a list from the Sustainable Agriculture Network (SAN) of the most popular cover crops in each region. Note that many of the species mentioned are also valuable as both hay and pasture plantings. I would add that you should compare seed prices, consider how hard a crop is to plow down with your equipment (some may have to be mowed and/or chopped before they'll till under well), and think about how hard a perennial crop will be to kill. Some will give up when you disc them, while others may have to be tilled under deeply.

NORTHEAST: Alfalfa, annual ryegrass, annual sweet clover, Austrian winter pea, barley, berseem clover, big flower vetch, blue lupine, bromegrass, buckwheat, cowpeas, crimson clover, crown vetch, fava bean, grain rye, hairy vetch, mammoth red clover, oats, red clover, subterranean clover, Sudan grass, wheat, white clover, white lupine, woollypod vetch, yellow-blossom sweet clover

NORTH CENTRAL: Alfalfa, annual ryegrass, Austrian winter peas, barley, big flower vetch, crimson clover, flax, grain rye, hairy vetch, red clover, spring oats, subterranean clover, sweet clover, wheat, white clover, yellow-blossom sweet clover

SOUTH: Annual ryegrass, Austrian winter peas, barley, berseem clover, big flower vetch, crimson clover, grain rye, hairy vetch, oats, subterranean clover, wheat

WEST: Alfalfa, Austrian winter peas, annual ryegrass, berseem clover, bluegrass, buckwheat, crimson clover, fava beans, grain rye, hairy vetch, medics, orchard grass, red clover, subterranean clover, white clover, yellow-blossom sweet clover

Beneficial Insect Plantings

Beneficial plantings can be designed to attract both pollinators (including honey bees) and insects that prey on insect pests, such as parasitical wasps. Both types of plantings can be integrated into the garden and orchard or planted around the edges as borders. For more extensive lists of plants specific to your region that attract beneficial insects, visit the websites of your area's organic farming and gardening associations and the state Cooperative Extension Service. The Xerces Society (see Resources for website) maintains region-specific lists of plants that benefit pollinators. Below is a starter list of plants used as beneficials in their flowering stage, culled from a 1990s issue of *New Farm* magazine. It features a number of plants more commonly thought of as weeds!

Widely Adapted Plants That Attract Beneficial Insects

PLANT	ATTRACTS	TO CONTROL
Berseem clover	Big-eyed bugs	A variety of insects and mites
Black locust	Lady beetles	Aphids
California lilac	Hoverflies	Aphids
Caraway	Lacewings, hoverflies, insidious flower bugs, spiders	Young corn earworms, aphids, thrips, and mites
	Parasitic wasps	Aphids, beetles, caterpillars, and flies, depending on species, are parasitized by wasp larvae
	Soft-winged flower beetles	Moth eggs
Common knotweed	Big-eyed bugs, hoverflies, parasitic wasps, soft-winged flower beetles	
Cowpea	Predatory wasps	Caterpillars; wasps rely on nectar so are attracted by flowering plants
Crimson clover	Minute pirate bugs	Insect eggs, thrips, mites, and larvae
Crimson clover (South)	Lady beetles	Corn earworm
Flowering buckwheat	Hoverflies, minute pirate bugs, predatory wasps, lacewings, lady beetles, tachinid flies	
Hairy vetch	Lady beetles, minute pirate bugs, predatory wasps	
Queen Anne's lace	Lacewings, predatory wasps, minute pirate bugs, tachinid flies	
Soap-bark tree	Hoverflies, green lacewings, brown lacewings	
Spearmint	Predatory wasps	
Sweet alyssum	Tachinid flies, hoverflies, chalcids	
Subterranean clover	Big-eyed bugs	A variety of insects and mites
Sweet fennel	Parasitic wasps, predatory wasps	
Toothpick ammi	Hoverflies, minute pirate bugs, soft-winged flower beetles, tachinid flies	
Tansy	Parasitic wasps, lady beetles, insidious flower bugs, lacewings	
White sweet clover	Tachinid flies, bees, predatory wasps	
Wild buckwheat	Hoverflies, minute pirate bugs, tachinid flies	
Yarrow	Lady beetles, parasitic wasps, bees	

Note: Adapted and reprinted with permission from the Rodale Institute (Rodaleinstitute.org) © 2016.

Trap Crops

Trap crops are a trickier thing to get right than beneficial plantings. They have to be specific to the pest and the main crop, and there is the potential to create pest nurseries if you don't control the problem insect when it invades your trap crop. Most sources recommend putting a lot of research and effort into planting appropriate trap crops only when there is a major and recurring insect pest problem. Trap crop plants can be positioned within the row of the main crop, in alternating rows or strips within the main crop, or as a border adjacent to the main crop.

Timing is important with trap crops. You can either put in an early planting of the main crop to attract the target pest and destroy it before it gets to the main crop, or you can seed a different variety or species that's more attractive to the pest than the main crop — as long as the trap crop will grow and flower at about the same rate. You may need to be creative about destroying the "trapped" pests in an organic system: some can be picked off by hand (Colorado potato beetles are a good example), some will vacuum off with a shop vac, and for some you may have to till under the entire crop.

A trap crop can minimize pest damage to your vegetables.

The following list of examples of trap crops comes from OISAT's Online Information Service for Non-chemical Pest Management in the Tropics (much of its information is derived from temperate zone research):

- Alternating rows of beans and corn to control stalk borers and fall armyworm
- Using collards as a border around cabbage to control diamondback moths
- Planting basil and marigolds to control thrips in garlic
- Alternating rows of dill and lovage with tomatoes to control tomato hornworm
- Interplanting horseradish and/or tansy with potatoes to control Colorado potato beetle
- Using hot cherry peppers as a border around bell peppers to control pepper maggot
- Planting marigolds in alternating rows with crops from the crucifer, legume, and cucurbit families to control nematodes
- Alternating rows of nasturtium with cabbage to control aphids, flea beetles, cucumber beetles, and squash vine borers
- Planting onions and garlic as borders around carrots to control carrot root fly and thrips

Topic of the Season

Honey Bee Partners

Honey bees are called domesticated, but they're not really. If we want to harvest honey, we have to adapt to their behaviors and schedule, because they won't be constrained by us. But they may be the most beneficial insects of all on our land, providing not only honey but essential pollination services for fruits and vegetables as well. When putting in plantings for beneficial insects, be sure to include flowers for the bees, especially species that will be in bloom when your fruits and vegetables are not.

Bee colonies contain a single queen who is fertilized by multiple drones, or male bees, in drone congregation areas. The drones that mate with the queen then die, and the queen spends the rest of her life laying eggs, fed and tended by worker bees, who are all female. The drone's only function is to mate with a queen; the worker bees variously collect pollen for feeding the young larval bees and nectar for making into honey to feed the rest of the colony. A strong colony in a good bloom year will make honey in excess of its needs, which can be harvested by the beekeeper.

In cold seasons when there are no blossoms, the bees must survive on stored honey and stay warm by collecting in a tight cluster inside the hive body, beating their wings to generate heat. If you open a hive in cold weather, you can kill the bees.

In late winter and early spring, the queen accelerates egg laying so that there are plenty of bees ready to collect nectar when flowers

Worker bees fill combs with honey.

open. It takes 21 days for a new egg to become an adult worker bee (drone eggs take 24 days). The demand for food for the increasing number of larvae can outrun availability at this time of year, so the hive may need feeding in the form of sugar syrup and/or solid pollen patties. Disease and pest problems also often become apparent at this season.

Colony collapse disorder is a relatively recent affliction of bees that is not entirely understood, but it is thought to be linked to exposure to pesticides. Bees may also suffer from American foulbrood, a bacterial disease for which there is no cure, as well as a selection of other bacterial and fungal diseases. They are also vulnerable to varroa mites and other pests such as chalk moths and mice. For this reason, most beekeepers routinely administer preventive medicines, such as antibiotics, and controls at specific times, such as entrance reducers to keep mice out in winter. The precise treatments and treatment schedule must be closely tuned to the weather and bee activity and thus will be specific to your locale. Visit or contact your state Cooperative Extension Service and area beekeepers' associations for the information you'll need. Some beekeepers choose not to use any treatments, but it's good to know what the potential problems and treatments are.

When setting up new hives, choose a secluded place where they'll have shelter from winds and light afternoon shade in summer. Face the opening to the southeast to catch the morning sun. If there's danger of raids by bears or other wild animals, put an electric fence around the bee yard. A little peanut butter on the wire to entice animals to lick it will make sure they receive the maximum amount of shock.

You can order hives, equipment (such as hats, suits, hive tool, and smoker), and bees from bee supply companies. The standard commercial hive is exactly built to suit a bee's very precise notions of how much room there should be between frames, which minimizes randomly built comb and brood cells and makes the honey harvest much simpler.

Honey is harvested in mid to late summer, and again in early fall if the hive has more than it needs for winter.

There are several classic texts on keeping bees, but it's easy to get bewildered by the detail. For a clear and comprehensible overview of the topic, I recommend *A Book of Bees: And How to Keep Them* (Ballantine Books, 1988) by author and commercial beekeeper Sue Hubbell. In it she writes, "The less I disrupt and fiddle with the bees, the more they can concentrate on making honey."

CHAPTER SIX

Early Summer

Begins after the average date of the last hard frost. Soil temperatures warm to 70°F/21°C or higher. First cutting of alfalfa is finished.

Summer begins when all danger of frost is past and the soil is warm enough to germinate anything. In any region where early summer brings both heat and humidity, insect populations are peaking at this time in barns, pastures, orchards, gardens, and fields. Stay on top of cleaning up manure in the barnyard and trimming weeds along the buildings to minimize pest bugs near the livestock.

Grass growth in hayfields and pastures is peaking or just past. You can judge its growth by walking your paddocks and fields regularly or by noting how often you're mowing the lawn. Though work in the garden and fields will continue, the highest priority this season is making the hay.

Early Summer

FOCUS ON:
making hay

Garden

» **Continue succession planting** and stay on top of the weeds.

» **Mulch if it's dry,** water as needed, and enjoy early vegetables and herbs. It's strawberry season.

Field

» **Make the first cutting of hay.**

» **Continue to cultivate row crops** as needed until the leaves of the plants in adjoining rows meet in the middle and shade the ground, suppressing weeds.

Pasture

» **Monitor paddocks** for slowing grass growth and adjust grazing rotation accordingly.

» **Continue weed control.**

» **Check temporary fences often** to make sure they're up and powered.

» **Mow fence lines.**

Orchard

- **Mow around trees.**
- **Thin fruits** when they're 1 to 1½ inches in diameter, if desired.
- **Clean up fruit** that drops to the ground.
- **Pick red currants,** gooseberries, and strawberries.

Beeyard

- **Add supers** as others fill up.
- **Take filled and capped frames** for honey.

Barn

- **Offer shelter** during the day if animals are badly bothered by flies. Young livestock should be getting the hang of the pasture rotation by now.

Coop

- **Watch for predator activity** and bring chickens closer to home if some start disappearing.
- **Order meat chicks** for midsummer delivery and fall butchering.

Equipment Shed

» **Pull out haying equipment** and make sure you have plenty of supplies and parts on hand.

Woodlot

» **Control invasive species;** there's not much else to do.

» **Cut timber and firewood,** if desired, and if it's dry and there are no disease concerns.

» **Mow the wood trails** if you're out mowing fence lines — this freshens up forages on the trails for grazing wildlife and minimizes ticks on trails.

Wildlife Habitat

» **Do a high mowing** of prairie plantings if annual weeds are overtaking the slower-growing perennial natives. Mow around new trees.

Seasonal Chores

Garden

Early in the morning, when there's still dew on the grass and you can't be in the hayfield, is the best time to get garden chores done at this season. There should be greens and some early vegetables on the table every day now, though it's not yet canning and freezing season. Stay on top of the weeds. With the vegetables well established, this is a good time to plant a living mulch between the rows. Or, if the weather is dry, put down a regular mulch — we use scraps of the just-cut hay for this, and it's a pleasure to work with such sweet-smelling material.

Field

Weed control is the focus at this season, though it can be hard to find time to cultivate when you're making hay. Because your crops are well established by now, it's a good opportunity (as in the garden) to plant a living mulch crop — especially a legume that will add nitrogen to the soil — before canopy closure shades the ground too densely.

Drying Herbs

Many herbs are now mature enough to harvest and dry for winter use, and, like mulching with hay, it's a sweet-smelling job. Wait until the dew has dried, then, using your fingers or scissors, remove the tops, leaving enough leaves to power regrowth. Most herbs can be recut regularly throughout the summer.

Herbs need a shady, clean, dry place to cure, and this can be difficult to find in summer in a humid climate. We tried the hayloft and screen porch over the years, with only mediocre results. A food dryer works, but hanging bunches on string in the basement near a dehumidifier has worked best of all. The results are herbs that are greener and more flavorful than store-bought products, and the bunches can hang all summer and be processed whenever you have time or need.

My favorite herbs to dry are basil, oregano, lemon balm, rosemary, and lavender. When dry, you can rub the bunches through a colander and put them in a jar in the spice cupboard, or just crumble leaves directly into whatever you're cooking.

Hang herbs to dry in small bunches somewhere cool, dry, and shady or dark.

Pasture

When weather patterns are within the normal range, the grass will continue to grow at a fast clip through this season. Cool-season grasses are beginning to slow down as the heat increases, while warm-season grasses are now starting to come on strong.

This is probably your last chance to control many pasture weeds before they set seed. By mowing or chopping them now, you also set them back harder, since they've poured most of their resources into making seed and their reserve capacity is at a low. But they will come back, so be ready to mow or chop again in late summer.

Orchard

If you want to thin the apple crop and didn't do it at blossom time, you have another opportunity now, when the fruits have grown to be about 1 to 1½ inches in diameter — big enough to see easily. The goal of thinning is to produce bigger and better-quality fruit this year, as well as a more uniform crop every year — especially with an apple variety that is a biennial bearer (one that will bear a good crop only every other year when left to its own devices).

Thinning should be done within 30 to 35 days of full bloom, according to *Growing Fruit in the Upper Midwest* (University of Minnesota Press, 1991), "to ensure flower buds for the next year and to have the greatest benefit on fruit size. Later thinning can be done to improve fruit size, but the benefit decreases as the growing season progresses." Thin to one fruit per bunch, with the fruits spaced no closer than a few inches.

Mow the orchard early in the season, so that dropped fruit will be easy to spot. Apple trees will "drop" some infected and/or overcrowded fruit on their own at this time, and you should clean up these drops to break up disease and pest life cycles. You can rake up

Young pigs will clean up June-dropped apples without damaging soil or trees, though you will need to keep an eye on them.

Weeds, Poisons, and Invasives in Pastures

Pasture weeds can be defined as plants that your livestock won't eat. Nonchemical control relies first on prevention, which includes having a healthy soil at the proper pH and encouraging a thick sod by rotational grazing. Unfortunately, that's usually not enough to discourage all of the weeds, so the second line of defense is mowing or hand chopping. This is best done when plants are in flower: when belowground root reserves are at a low and before the plants set seed. Many weeds can set seed within days, so it's better to be a bit early with this chore than a little late.

One of the most widespread and detested weeds in the United States and Canada is Canada thistle, a perennial variety of thistle that often eludes the most persistent efforts to eliminate it. Donkeys will eat it if they're hungry enough, but regular mowing might be simpler. If you don't get to mowing it down and it does go to seed, the goldfinches at least will be happy — thistle seed is one of their favorites.

In addition to plain old weeds, some plants can sicken or kill your livestock. Since there are different poisonous plants in different parts of the country, ask your state Cooperative Extension Service for a complete list for your area. Some of the most widespread are jimsonweed, buttercup, black nightshade, black locust, curly dock, locoweed, lupine, pigweed, and white snakeroot. Poisonous plants should be uprooted.

Invasives are nonnative species that will completely displace forages. Some aggressive native species will do the same thing. Burdock, for example, is a common native plant, but it will crowd out grass if it finds the right growing conditions. Mowing is not very effective, since burdock regrows so quickly; the best nonchemical way we've found to kill it is by cutting off the root below the soil line with a sharp shovel. Purple-spotted knapweed is another widespread and especially troublesome species. If you have too little time and too much weed to uproot, keep mowing it so it doesn't set seed.

Thistles may have food value for some wildlife, but you'll want to minimize their presence in your pastures.

drops by hand or put up a little temporary fence and let some small livestock in — sheep, goats, or young pigs — to do the job for you. But don't use grown pigs or cattle for this job; they will damage the trees.

Beeyard

The big spring blossoming has wound down and the summer flowers are coming in, though how fast and how many depends a lot on the weather. If there's an interval between spring and summer nectar flows, then leave honey in the hive. If conditions are good, with plenty of blooms available and fine weather, then supers that have been filled and capped can be removed for extracting the honey. Keep adding empty supers as needed.

Barn and Coop

By now, mothers and babies should be moving fairly smoothly through the pasture rotation. Aside from daily checks and rounding up the occasional kid that gets on the wrong side of the fence, the only other job is to continue to keep flies and other insect pests under control.

Since bugs thrive where there's moisture and cover, your primary means of control is to eliminate this type of habitat. Keep cleaning up manure in the barnyard and trim back grass and brush along fences and buildings. By putting the water tank on cement, you can eliminate a lot of insect-attracting mud and puddles. A drainage ditch at the low end of the barnyard does the same. When flies are at their worst, let animals into the barn or provide an open-sided shed for livestock to use during the day — most flies dislike the shade.

Tracking Nectar Flows

Bees rely on a wide variety of flowers to fill their honeycomb; just which ones depends on where in the country you're keeping hives. State and local beekeeping associations are good sources of information, but the only way to really know what's going on is to learn to identify the flowers that attract your bees and to keep a calendar of when they start and finish blooming. This will alert you to gaps in the nectar and pollen supply during the growing season.

Shearing Sheep

Sheep are shorn of their wool in late spring into early summer (hair sheep do not need shearing), either by the owner who has taken a class or by a professional sheep shearer. Professional shearers tend to travel long distances and be in an area only for a short while, so call early in spring to set up a date.

Selling the wool is problematic at best in the United States, and at worst impossible, especially if it's dirty or full of burrs. Keeping sheep on clean pastures and out of the mud makes all the difference.

If there are few or no sheep shearers in your area, and you have a strong back, consider learning the art and turning it into a part-time seasonal business.

Equipment Shed

It is crucial that you keep equipment working during hay making, since in so much of the country you're racing to beat the rain, and a breakdown that can't be quickly fixed may mean putting up rained-on hay. It's not the end of the world, but it's not good, either: rained-on hay is less palatable and has less nutritional value and poses an increased risk of becoming hot enough to cause a fire.

Before you start cutting, make sure you have supplies and parts on hand. These includes grease, engine oil, oil for lubricating chains and other moving parts, cotter pins, grease fittings (zerks), spare belts, chain links, and whatever else your equipment tends to wear out or break. We put these and selected tools (including a grease gun and a set of socket wrenches) in the back of our six-wheeler and park it out in the field so that we can drive tools and parts to the equipment instead of having to bring the equipment back to the farmyard. For small breakdowns, this can save quite a bit of time.

Woodlot

Nothing needs to be done in the woodlot at this season, but if you don't have field and garden work to keep you busy, you can mow trails and remove invasive plants. If it's dry and there are no disease issues in your area (for example, we don't cut oak when it's above 40°F/4°C degrees, to prevent oak wilt infection), then you can also make firewood and cut timber.

Wildlife Habitat

If you planted trees or shrubs in the spring of this year or the one previous, check them to make sure they're not being overtaken by the grass. If so, then mow around the saplings (mulching at planting time helps a lot, but often it is not practical when you're putting in a lot of trees or shrubs). Do the same with prairie plantings and other native perennials — these are often much slower growing than weeds, and a high mowing will set the weeds back without harming the smaller natives.

Making Hay

Hay consists of grasses and forages cut while still green and growing, then dried and stored to feed livestock when the growing season is over. Since hay is the main component of the livestock diet on a small-scale operation, making it yourself (or at least having someone else make it from your fields) saves a lot of money on the feed bill. On the other hand, for anything more than a few acres, making hay requires some fairly serious equipment and significant time. With just a little patch, you can make hay with a scythe and a rake if you have a strong back.

There are five steps to making hay: cutting, drying, raking, gathering, and storing. But before starting, you'll have to make a judgment call on when the crop is at the right maturity and the weather is likely to be dry and sunny for four or five days — and those two things don't always coincide.

Maturity determines palatability and nutritional value: The more mature the plants, the less nutritive and palatable, so the less your livestock will benefit. On the other hand, if you cut the stand when it is too immature, you can slow its regrowth and your yield will be lower. Young hay, like young pasture — especially if it's pure alfalfa — can be excellent for producing milk and growth but hard on digestive systems and too fattening for animals that aren't milking or growing. For most small-scale operations, it is fine to cut alfalfa and clovers just as they're blooming and to cut grasses before seed heads are fully mature. A mixed hay of medium maturity is a good middle ground.

To make small square hay bales, you need a crew and weather-tight storage, but the result can be very high-quality hay.

It usually takes three to five days to make hay: a day to cut, a day or two to dry, and a day or two to rake and bale or fork onto a wagon and place under cover. Ideally, the weather will be dry this entire time, but if you can't avoid rain entirely, at least wait to rake and bale the hay until it is dry. When you can pick up a hank and it feels dry but isn't so dry it crumbles, that's the time to rake. Wet hay raked into a windrow will not dry reliably and will at best mold in patches, at worst heat and start a fire. Once the hay is raked into windrows, you don't want to bale it until it's crackly-dry. This may take an hour in really hot, dry weather or a day if it's quite humid. Don't just test the tops of the windrows; kick them over here and there and feel the hay underneath, too.

You can cut hay with a scythe or mechanical mower — anything from a sickle bar cutter to a haybine to a discbine and variations on these basic implements. Generally, you first cut several rows around the outside of the field, then cut up and down the interior of the field, working in blocks.

A scythe is an ancient tool that still works well in a small field.

Rake the outside rows so they're not so close to the fence or field edge that the baler can't pick them up, and not so close to the interior rows that you won't be able to turn the tractor and rake without disturbing the edge rows. How exactly you do this will depend on your equipment. Small walk-behind equipment can turn in a small area; our tractor with the V-rake cannot. If you're making small square bales or picking up the loose hay, rake each cut row into its own windrow. For a round baler or large square baler, you can rake several rows together.

Hay can be gathered with a pitchfork and flatbed trailer or even a pickup truck, a square baler and hay wagon, or a round baler. Large square balers are generally beyond the needs or budget of a small farm, and the bales are difficult to move without a skid-steer loader and big flatbed trailer. We originally made small squares, then in subsequent years we paid a custom operator to round bale for us after I'd done the mowing and raking, and now we use a round baler we bought at an auction. This has made haying much quicker than it used to be, but it's also a significant investment in equipment — one that we couldn't have afforded to make except over many years.

Hay that's gathered loose or made into small square bales should be stored under cover: either in a barn hayloft or in a shed with a floor that won't transmit moisture.

This means putting a solid plastic tarp or spaced boards under the bottom layer of hay. Round bales can be stored outside. Rain and weather may spoil the hay a few inches deep, but the inside of a well-made bale will stay good for a couple of years at least. If you can put round bales under cover, they will last longer. The spoiled hay on the outside of the bales doesn't go to waste; it can serve as bedding for the cattle, mulch for the garden, or organic matter for the soil. (*Note:* Round bales are bound with either twine, plastic netting, or plastic wrap. Netting and wrap may preserve more hay but are harder to remove and create a big disposal problem.)

Haying Equipment

Haying equipment has evolved steadily over the past two centuries, but seemingly every tool or implement that's been used to make hay in the past is still available today, if you look hard enough and have the mechanical (and often welding) skills to get antiques running again. Cutting implements begin with the scythe and go through horse-drawn and tractor-powered sickle bar mowers to haybines (sickle bars with crimping rolls that crush the hay slightly so it dries faster). The discbine uses rotating blades (somewhat like a lawnmower) with rollers. It is by far the quickest way to cut hay, and it requires fewer repairs. It's also quite expensive. Walk-behind haying equipment is starting to show up in this country, too, and for an area too big to scythe but too small to justify investing in tractor-powered implements, it might be just the ticket — if you have the budget.

Traditional wooden hay rakes were succeeded by horse-drawn dump rakes, side-delivery rakes, and rotary and pinwheel rakes. My favorite is still the side-delivery rake — with a dolly wheel on the front, it's a snap to hook up, and maintenance and repairs are generally obvious and simple. But the rake

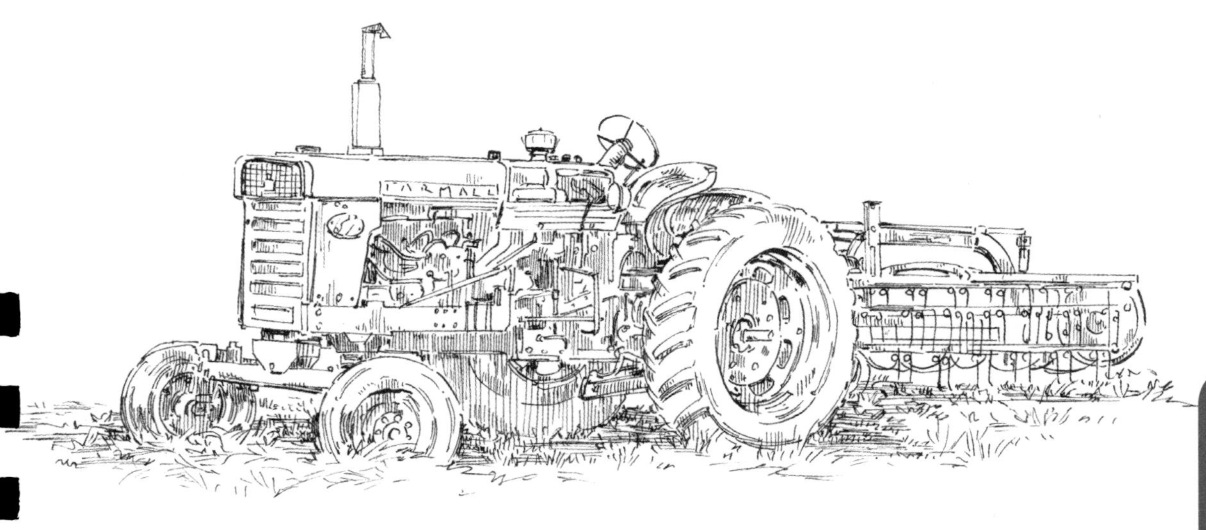

A side-delivery rake is simple to run and easy to fix.

What to Grow for Hay

If you have a field that's already growing grass, and maybe even some clovers, that is a good place to start making hay. If your acres are weedy, have been in row crops, or brush has moved in, you may want to plow it down and replant. A mixed hay of grass and clovers or alfalfa will generally last the longest without needing replanting, and it will furnish a more balanced diet for your livestock than either pure grass or pure alfalfa. But do what you think will work best on your land. Below are listed some of the most common plants used for hay.

GRASSES

Bermuda grass: Adapted to the Deep South, Bermuda grass should be burned each spring to maintain the stand. Use a hybrid variety adapted for hay.

Smooth bromegrass: It is best adapted to moderate rainfall and moderate summer temperatures.

Orchard grass: Widely adapted bunch-type grass, except in dry regions; orchard grass is not as cold-tolerant as brome.

Timothy: Adapted to cool, humid climates and very winter hardy, timothy doesn't tolerate heat or drought well. It is a personal favorite.

Others: Other grass species used more regionally for hay include big and little bluestem, various wheatgrass species, tall fescue, quack grass, redtop, and switchgrass.

LEGUMES

Alfalfa: The most widely grown of all the legumes, alfalfa is popular for its high yields and palatability. It doesn't do well in the Deep South or on heavy soils.

Red clover: It is best adapted to northern states. It tolerates heavier and more acid soils than alfalfa and makes very palatable hay, though it's slower to dry in the field.

Crimson clover: It is similar to red clover, but better adapted to southern areas.

Lespedeza: It is better adapted to acidic, sandy soils than other legumes, and it tolerates heat but not cold.

A hay crop can be planted either in early spring or in late summer. If planting a pure stand of legumes, consider mixing it with a small grain, such as oats, as a nurse crop. Though this practice is no longer common, it still works well in an organic system for suppressing weeds and providing winter cover. Better yet is a mixed planting of grasses and legumes: the stand will endure longer, have few problems with leafhoppers and other pests, and give your livestock a more balanced diet.

we use most of the time now is the pinwheel V-rake, since it's three times as fast. For a few acres, a side-delivery rake is reasonably priced (used) and can be pulled behind a pickup or even an ATV or a tractor.

You can gather hay with hay forks (pitchforks with many thin tines), a hay loader, a square baler, or a round baler. A round baler is a one-person operation; the other methods are best if you have plenty of help.

This discussion by no means covers all the details of making hay, especially when you consider that every region has its own quirks when it comes to summer weather and preferences for equipment and plants. Talk with the neighbors to pick up on the details, or better yet, help a neighbor make hay.

Cutting Hay and Grassland Birds

Several species of birds rely on open grasslands for nesting sites. Unfortunately, many of these species are dwindling as pastures and hayfields have been replaced by row crops and development, and especially as the remaining hayfields are cut much earlier in the season — during the middle of nesting. Alfalfa in particular has replaced grasses and clovers for hay in much of the country, and it is generally cut in late spring. If you have nesting populations of bobolinks, meadowlarks, killdeer, or other grassland birds in your hayfields (we have bobolinks), consider waiting to cut your hay until you see the fledglings sitting on fence wires.

CHAPTER SEVEN

Midsummer

Cool-season grass growth slows (when daytime air temperatures are over 75°F/24°C) or stops completely (when temperatures are over 90°F/32°C). Warm-season grasses grow rapidly (best at 80 to 90°F/27 to 32°C).

IN THIS, THE HOTTEST of seasons, try to concentrate outdoors work in the morning and early evening hours when the temperature is more tolerable, spending mornings in the garden and early evenings clipping paddocks or fence lines. Harvest season really begins now, with honey, small grains, meat chickens (if they were started in spring), and most of all the increasing flow of produce from the garden. A lot of time in midsummer is spent in the kitchen canning, freezing, pickling, and drying.

On the other hand, field crops and garden vegetables are getting high enough to shade weeds, livestock and poultry are happy on pasture, the first crop of hay is done, and fair season is getting under way. This is a nice opportunity to relax a little and see what your neighbors have been doing on their places.

Midsummer

FOCUS ON:
garden

Garden

» **Keep watering.** Weed pressure lessens and harvesting accelerates.

» **Sow fall crops,** such as garlic, carrots, parsnips, rutabaga, cabbage, kohlrabi, and salad greens — anything that will mature in half a growing season and tolerate cool fall weather — as vegetables are harvested.

» **Test your soil.**

Field

» **Harvest small grains.**

» **Stop cultivating;** row crops should be high enough by now to shade weeds.

» **Test your soil.**

» **Spread manure and lime** on cut hayfields.

Pasture

» **Slow down rotations** to give paddocks adequate recovery time, or add paddocks, or shift the rotation to paddocks with warm-season forages. As the weather gets hotter and dryer, growth of cool-season forages will continue to slow.

» **Clip paddocks after grazing,** if needed, to control weeds.

» **Test your soil.**

Midsummer

Orchard

- **Continue to monitor** for pests and disease.
- **Test your soil.**
- **Harvest peaches,** plums, apricots, both sweet and sour cherries, raspberries, and huckleberries.

Beeyard

- **Begin honey harvest;** leave a full super for bee food.
- **Supply hives** with more empty supers.

Barn

- **Cull infertile or problem animals.**
- **Start meat animals** on grain if desired.
- **Turn out the bull.**
- **Make sure shade or shelter is available** on very hot days.
- **Let animals in** during the day and out to graze at night if flies are really bad.

Coop

- **Butcher meat chickens** started in spring.
- **Start chicks now** if you intend to butcher chickens in fall.

Equipment Shed

- **Clean up haying equipment** and get it ready for the next cutting.
- **Pull out implements** for small grain harvest.
- **Keep mower blades sharp.**

Woodlot

- **Cut timber for firewood** if it's dry and you have the time. Though if flies and heat are intense, why not wait for cooler weather?

Wildlife Habitat

- **Place turtle logs** and fish cribs.
- **Put coarse rock** in shallows for small fish.

Seasonal Chores

Garden

The garden is in its glory by midsummer. Harvest volume accelerates, and it's a rush to get everything picked, processed, and stored. As the weather continues to get drier and warmer, maintain mulch and/or cover crops. Keep watering.

As early crops like cabbage and peas are harvested, plant something to replace them: maybe a green manure that can be tilled down when the weather cools off enough to put in more peas and cabbage or some fast-growing root crops for fall harvest. This is a good time of year to test your soil.

Field

With canopy closure, row crops should now be fine on their own until harvest, but small grains are ripening fast. They're ready to harvest when the grain is brittle hard (it cracks when you bite it) and the stalks are fully brown. Ideally you'll cut the grain slightly before or just as it reaches this stage, before birds attack the crop or the ripe grain starts falling to the ground.

The average small holder doesn't own grain harvesting equipment, either antique or modern (swather and thresher or a combine), and it would be a rare thing to have a neighbor with a combine willing to come over for a little field or a garden patch of grain. For most of us, a sickle or scythe is the most practical harvest method. When cutting, be careful to lay all of the grain the same way, so that all of the seed heads are at one end. Place the crop under cover, spreading it to dry or tying it in bundles and hanging them or leaning a group together and tying it so it will stand up with seed heads at the top. The grain can remain there until you're ready to thresh it (separate the grain from the stalk and hull).

Threshing can be done a few different ways, depending on the grain and the variety.

Putting Up Food

Garden and orchard produce can be preserved by home canning, pickling, drying, freezing, or storing in a cold cellar, according to the vegetable and your personal preferences. It's best to store your vegetables as quickly as possible to produce a better-quality product, but sometimes you're a little short of both time and storage space.

There are useful shortcuts to take the pressure off in a busy season. For example, if your cold cellar isn't cold enough during summer and early fall, you can use a second refrigerator instead, or freeze or can some of those vegetables — I've found that cabbage and root crops freeze well if you chop them and spread them in a pan to put in the freezer. Once frozen, bag them up. If you have a lot of fruit and not enough time to convert it into jams and jellies, you can make the juice and freeze it until sometime next fall or winter and can the jelly then.

You can tie wheat in sheaves and store for later threshing.

For grains with seed that will fall off easily, holding a bundle and beating the seed heads with a stick or against a wall will do the job. (The traditional tool is a flail, two sections of wooden rod joined by a very short length of chain.) Grain that is tight on the stalk, like wheat, takes a little more work. The best method we've found mimics the traditional one of having donkeys or oxen tread on the crop: we spread the stalks on a rough wooden floor and dance on them (doing the twist is the most effective step). Then we pour the grain back and forth with two buckets in front of a fan a few times to blow away the chaff. For a bigger crop, it would make sense to invest in a small-scale mechanical thresher.

Once the grain has been harvested, plant a green manure or a forage crop for hay or pasture. Plant a different small grain to insulate the young forage legumes over the winter (oats with alfalfa is a good combination) and minimize weed competition.

If soil test results recommend you add amendments to the hayfields, add them now before the second crop gets too high. Rain will move the amendments into the soil, where they will be quickly utilized this time of year, when soil life is most active.

Pasture

Continue adjusting the rotation to the rate of grass growth and cutting weeds. In an average year, the weather is at its driest and hottest from now through late summer, and this slows forage growth. To avoid grazing the grass too short (which greatly slows recovery time), feed hay or add paddocks. You can use part or all of your hay ground for this, but it will mean less second crop hay: you're making a choice between more feed now or more feed this winter.

Another way to take the pressure off of pastures is to do a midsummer cull: send females that didn't reproduce, or that have difficult temperaments, off to the processer or the auction barn. This leaves more grass for those animals that are fattening or reproducing or milking.

Orchard

Summer fruits are ripening now, first those on the outer and upper branches. Enjoy them fresh or make jelly, jam, fruit butter, preserves, or juice. If any poultry, livestock, or wildlife (usually deer) have been in the orchard, do not use fruit that has been on the ground, since it could be contaminated by pathogens found in animal manure. These drops should be picked up and can be fed to the livestock, left in the woods for wildlife, or composted.

Beeyard

This is primary honey harvest time, as honey production slows. When removing filled supers for extraction, be sure to leave a full super of honey for the bees (especially the most vigorous hives), who will need it for food as blossoms dwindle at this time of year.

Barn

This is a good time to cull animals that you aren't fattening for market and don't wish to carry through winter — those with bad dispositions or that are no longer bearing young. This leaves more pasture for the animals you do want to keep at a time when pasture is usually getting a little short.

If you haven't been feeding grain and wish to put a "grain finish" on livestock intended for market or freezer, this is a good time to slowly add grain to the daily ration. Grain can be started anytime from before weaning to just a few weeks before butchering, or never introduced at all. Here are a few thoughts to help you figure out what to do on your place.

- The natural diet of cattle and sheep does not include grain; it's all forages. One hundred percent grass-fed lamb and beef can taste fabulous — or awful. To do grass-fed right, you need good genetics (animals that are genetically predisposed to gain weight rapidly on an all-forage diet), great pastures, and excellent pasture management.

- Starting grain a week or two before weaning serves a number of purposes: it takes pressure off the mothers, it gives the youngsters a boost ahead of the trauma of weaning, and it's an easy way to train the young to come into the pens or the barnyard willingly. We also use it as a

Turn the bull out with the cows nine months before you want to see the first calves. Be sure to move heifers too young to be bred but old enough to be cycling well away from the bull.

training aid: the feed bunkers for our cattle are along one side of the holding area for the chute and headgate we use when we need to vaccinate, castrate, load, or sort. After the calves have had a few days to get used to the idea that there are treats in the holding pen, I shut the gate behind them and open the chute. In a couple of weeks, they're used to exiting through the chute, which makes them much calmer and easier to work with. On the other hand, starting grain that early means a higher overall feed bill, or a bigger grain field to harvest.

- Starting grain in late summer, a few months before sale or slaughter, as pastures slow down, is a very effective way of adding desirable fat and muscle. There is adequate time to get animals trained to come when they're called and be comfortable with handling facilities.

- Starting grain two or three weeks before sale or slaughter will add some weight for the least expense, but it will not improve tenderness or taste nearly as much as starting earlier.

For calves in mid-spring, cattle raisers turn the bull in with the cows at this time. For a few animals on a small acreage, it works well to buy a yearling bull and leave him with the cows until the following summer, then sell him. A young bull is easier to handle and will breed the cows twice and be gone before he is old enough to "get an attitude" or his daughters are old enough to be bred by their sire.

Coop

If you started meat chickens this early spring, they're about ready to butcher by now. They should be butchered early in the morning, before the heat wakes up all of the flies. Or you may prefer to start your meat chickens now, to butcher in early to mid-fall, when it's cooler. If you're marketing home-grown pastured poultry, you may be butchering every two weeks from early summer to mid-fall.

If you plan to butcher chickens in mid to late fall, start the chicks now, 8 to 12 weeks ahead of your planned processing date. Instead of protecting the chicks from cool weather, as was necessary in spring, now you have to pay attention to the heat — if they get too warm, they will pile in the corners of the brooder and the ones on the bottom will suffocate. If necessary, you could even put them in your cool basement for a couple of weeks until they're better able to moderate their body temperatures.

Equipment Shed

After you've done the first cutting of hay, get the equipment ready for the next one: blow off or power wash (never power wash the baler — blow it off), lubricate, check tightness and wear of belts and chains, check tire pressures, and look for leaks and loose bolts or other parts.

How to Butcher a Chicken

The evening before the big day, take away the feed so the chickens' crops are empty by morning (this saves some mess if you accidentally cut into the crop). After dark, if they're not in pasture pens, catch the chickens and confine them in a cage or other small area, so you can catch them in the morning without a lot of fuss. The goal is to keep everything as quiet and smooth as possible; you don't want to chase frantic birds around a coop or yard.

Kill the birds by chopping off their heads with a hatchet or by putting them upside down into open-ended cones and cutting their throats. Either way is quite quick and simple. (There are two good ways to get the bird to hold still before you bring down the hatchet. Either smack it hard on the back of the head with a heavy stick, or pound two nails into a sturdy board, as far apart as the width of a chicken's neck. Put the neck between the nails, then grab the feet and pull the head up snug to the nails. This makes it much easier to hit your target.)

Next, dip the carcass in water that is just under the boiling point (around 200°F/95°C) until the feathers loosen (we use an outdoor gas burner with a big pot on it). Pull out the carcass and strip the feathers — you can get quite fast doing this by hand, but if I were ever to go into the chicken-selling business, I would build or buy a mechanical plucker.

Once plucked, lay the bird on a table, open the rear with a sharp knife, and extract the innards. Then wash the carcass, bag it, and put it in the fridge or freezer. There are plenty of little details to know about this process in order to do it efficiently and without contaminating the meat with the contents of the digestive tract. I highly recommend getting a detailed book or manual, or helping someone else several times before you process your own birds. I learned from Wally, a nice old retired farmer who used to live a couple of miles down the road.

Woodlot

In regions where wet winters make woods work damaging to soils and dangerous to people, the dry days of summer are an opportunity for harvesting and improvement projects. Have a jug of water along in case of sparks — dry weather always represents a fire danger. Also, a county forester or woodland owners organization can advise you on which species of trees are threatened by insects and airborne diseases in your area. These should not be pruned or cut except in cold (below 40°F/4°C) weather.

Wildlife Habitat

Turtles like to sun themselves on warm days, but they need to get into the water quickly if a predator appears. A log with one end on shore and most of it extending into the water is a favorite spot, since it offers a clear view around and a quick exit. Drop a couple of such logs in your pond or stream.

If you have a pond, lake, or stream with fish, protect the young fish from older ones by adding coarse rock in the shallows or fish cribs in deeper water. The nooks and crannies in these structures are just the right size for smaller fish, and the big ones can't get in. In winter you can position fish cribs on the ice, where they'll drop into place at ice melt. In summer the cribs have to be towed by boat into position, using whatever floatation you can devise, and then sunk. As noted in chapter 1, crib placement is regulated in most areas; contact your state Department of Natural Resources for information before undertaking this project.

Topic of the Season

Caring for an Orchard

Unless you plan to sell fruit, one to three trees (standard sized or semi-dwarfed) of any one type of fruit will probably provide plenty for home use and even enough to give some away in good years. If you start trees in suitable soil and terrain, space them for good airflow and plenty of sun, and keep them pruned, you usually won't need to do much more except enjoy the harvest. There will be damaged fruit from insects and diseases, but most years there will be enough good fruit for fresh eating, and the rest can go into jams, jellies, pies, and juice. If you are interested in being more active in controlling pests (which might be necessary if you're in a major fruit-growing area where pests are prevalent), I give general guidelines for orchards below. Also read about integrated pest management (IPM), page 141.

An established orchard can provide a lot of food — as well as a lot of beauty — for not too much effort.

Fruits

Besides tree fruits — apple, apricot, cherry, pear, plum, and others — many other fruits can be grown in the garden or, especially with thorny or woody types, in separate patches. They're all a great addition to a rural home. The dates below will vary considerably by weather and variety grown.

APPROXIMATE ORDER OF FRUIT RIPENING

Approximate order of fruit ripening
Late winter in subtropical climates: Oranges, tangerines, lemons
Spring to early summer in subtropical climates: Grapefruit
Mid-spring: Rhubarb
Early summer: Red currants, gooseberries, and strawberries
Early to late summer: Plums, peaches, huckleberries
Midsummer: Apricots, summer-bearing raspberries, blackberries
Late summer: Blueberries, early apples
Late summer to early fall: Grapes, pawpaws, pears (*Note:* Pears are picked when firm, but not ripe, and ripened indoors)
Early fall: Some apples, fall-bearing raspberries
Mid-fall: Mid- to late-season apples, cranberries

Varieties of many types of fruit are adapted to a wide range of growing zones, so most home orchards should be able to produce a nice selection of fruit. We have apples, plums, apricots (though we're so far north we get fruit only about two years in five), and cherries in our orchard, with red currant bushes and Juneberries along one border. All of the fruit trees are planted and pruned in the same way, for the most part, and there's even overlap in the pests and diseases that bother fruits. Where specific examples are given in this book to illustrate a point, I will use apples, the most widely grown of the tree fruits.

A healthy orchard begins with fruit trees that you have selected for their hardiness in your climate and resistance to regionally prevalent diseases (for example, for apples in our area, the main concerns are fire blight and apple scab). You can easily find information on the pests and diseases in your area from state Cooperative Extension Services and fruit growers associations. Also choose varieties that have the characteristics you want: early or late fruit, dwarfed or full-sized trees, and fruit that is good for your intended purposes — such as fresh eating, juice, or other products.

Fruit trees prefer a loamy, moderately fertile soil on a slope that will let cold air drain away. If late spring frosts are not uncommon during regular bloom time, you can plant the trees on a north-facing slope to help delay flowering. Test the soil before planting, and work in any recommended amendments.

Space the trees according to their expected spread at maturity — for full-sized apple trees, 30 feet is about right. You want them to have at least a few feet of open space in between. Placing them in rows makes moving

Pollination Requirements of Fruit Trees

Tree fruits are either self-fertile, meaning that the pollen from a tree can successfully fertilize its own flowers, or they need to be cross-pollinated with pollen from another tree to have successful fertilization. For fruit trees that are not self-fertile, plant at least two trees of the same variety or of different varieties that will bloom at the same time.

Self-Fertile	Require Two or More Plants
Peaches (most varieties)	Apples
European plums	Pears
Apricots	Japanese plums (most varieties)
Figs	Cherries (most varieties)
Grapes	Blueberries (plant different varieties)
Persimmon	Nuts
Berries, except for blueberries	

equipment, such as mowers or a truck to carry pruned branches or fruit, easy and efficient.

Then provide habitat for insect-eating birds and beneficial insects like parasitic wasps that will help control insect pests common in orchards (see page 83). This will make a big difference but not solve all potential problems. For apple trees in the Midwest, a beneficial habitat will probably keep the worst pests — codling moths, plum curculios, and apple maggots — under control if they are prevalent in your area. More on this later.

Whether or not you cover the ground under and around the trees is an individual preference. There are no perfect solutions: any ground cover will aid some pests and help control others, and all of the options have pluses and minuses regarding soil health, rodent control, sanitation, and maintaining pollinators and other beneficial insects and birds. The options are:

- **BARE GROUND AROUND TREES.** This option improves effectiveness of sprays, encourages erosion, makes some orchard work muddy, and minimizes habitat for all insects.

- **SHORT GROUND COVER.** This makes it easy to clean up drops and prunings, and it limits rodent populations and many insects. It discourages beneficial insects that need flowering plants and cover for parts of their life cycles.

- **LONG GROUND COVER.** A longer cover makes it difficult to clean up drops and spot problems, and it can provide excellent habitat for beneficial insects and cover for small birds (good) and rodents (bad).

- **MIX OF SHORT AND LONG COVER THROUGH THE YEAR.** This is our compromise. We mow until apple drop has occurred, then we let the grass grow long until mowing again just before harvest begins. In mid to late fall we mow again so it's short for winter. There are more rodents in summer as a result, but very few in winter when they're most likely to damage bark on young trees.

Planting New Trees

To give fruit trees the best start and reduce labor required through the year, put the trees in the ground as soon as it thaws. Though if you pay extra attention to water, you can plant anytime up until early fall.

To plant a fruit tree, dig a hole a little wider and deeper than the sapling's root system. Loosen the soil at the bottom of the hole and throw in a little compost or the upside-down pieces of sod from the top of the hole. Carefully spread and straighten the roots, put them in the hole, and replace the soil. Add water when the hole is half full, and again when all of the soil is replaced, to settle the soil and remove any big air spaces (these will kill roots). The soil surface should be just below the graft line on the trunk and slightly below the surrounding soil to encourage water to pool and sink in. Mulch heavily around the tree, but leave a couple of inches around the trunk mulch-free to prevent mold and discourage insects. Place a 12-inch-high cylinder of screening or hardware cloth firmly in the soil and loosely around the trunk to prevent rodent damage (you can also use purchased plastic tree tubes, though they're not my preference). To prevent deer damage, put electric fence around the orchard or a 6-foot cylinder of woven wire around the tree. In mid-fall, paint the trunks with a 50-50 mix of water and white latex paint to prevent sunscald (cracking of the bark due to being heated up by morning sun too quickly after cold nights) during winter.

Pruning

Prune every year or two to open up the tree's structure so that sun and air can easily penetrate, giving you better fruit and healthier trees. Pruning is traditionally done in late winter, when the trees are bare and it's easy to see where branches are crowded. (For a discussion of how to prune, see page 28.)

Sanitation

"Sanitation" means cleaning up pruned branches and fallen apples to break up insect and disease cycles. Apple scab fungus, for example, overwinters in infected leaves on the ground; if this is a concern in your orchard, then clean up the leaves by mowing or shredding if there are too many to rake. Plum curculio larvae live in many of those June dropped apples, so clean up the drops — you can rake them up and destroy them (don't put them in compost) or turn sheep or young pigs out to do the job.

Active Pest Control

If you plan to be more aggressive about pest and disease control in the orchard, the first step is to identify what you find or expect you might find. Then research the life cycles of these pests and keep track of degree days (see Degree Days, below). Put controls in place just before or as the number of degree days adds up to a hatch or a weak point in a life cycle. For example, hanging out white blossom–mimicking sticky traps at blossom time will attract and trap sawflies, while plum curculios may be controlled by pheromone-baited sticky traps during their mating season or discouraged by coating apples with a kaolin clay compound.

Degree Days

Both insect and disease cycles are highly correlated with the accumulation of degree days, a measure of the total amount of heat needed for an organism (including bacteria and viruses) to move through its life cycle. A simple method to track degree days is to use a "high-low" thermometer that records the daily high and the daily low. Add these numbers together and divide by two to get the average temperature for the day. Subtract 50 to get the degree days accumulated for that date (negative degree days don't count). Keep a running total and correlate it with the degree day requirements for the life cycle of the pest you're targeting. Note that degree day requirements for the same pest will vary from region to region, so when you find information on the Internet, make sure it's applicable to where you live. For example, the first emergence of adult apple maggots occurs at 1456 degree days (Fahrenheit) in Provo, Utah; at 1129 degree days in Geneva, New York; and at 1148 in Guelph, Ontario.

CHAPTER EIGHT

Late Summer

Daily high air temperatures begin to cool off and cool-season grasses begin to grow again. Warm-season grass growth slows.

IT'S STILL HOT, but the days are getting noticeably shorter and perhaps even cooling off slightly as fall approaches. Produce continues to stream into the kitchen, and it's time to cut hay again. The second cutting of hay is more relaxed than the first: the urgency is gone. The second crop (and subsequent cuttings if you make them) can be left to "stretch" for a couple of weeks without becoming too fibrous. The weather is usually more cooperative this time of year and the hay crop is lighter, so it's easier on the equipment and quicker to dry, rake, and bale.

In an average year, this will be the driest season, making it the right time to dig wells, ponds, and drainage ditches and install culverts. If you have drainage ditches already, this is the best time to clean them out.

Late Summer

FOCUS ON:
garden • making hay

Garden

» **Continue harvesting your crops.**

» **Sow the last succession crops** such as spinach and lettuce.

Field

» **Cut your second crop of hay.**

» **Try a "stale bed" method** of weed control in the fields where small grains had been before planting forages or winter cover crops, if weeds were a problem.

Pasture

» **Adjust rotation as appropriate** for rate of grass growth.

» **Continue weed control.**

Orchard

- **Mow around trees.**
- **Harvest ripe fruit.**
- **Pick blueberries,** black currants, pears, early apples, and early grapes.

Beeyard

- **Finish the first honey harvest.**
- **Add supers** as needed.

Barn

- **Ready facilities** for weaning and other upcoming stock work.

Coop

- **Expect egg laying to stop** as older hens are molting.
- **Place the new batch of meat chickens** out on pasture.

Late Summer

Equipment Shed

» **Keep haying equipment running.**

Woodlot

» **Upgrade wetland crossings** on woodland trails if the weather is dry.

Wildlife Habitat

» **Dig a pond** if desired and the weather is dry.

» **Mow forage food plots.**

Garden

Produce continues to pour from the garden at this season, and kitchen activity is in high gear. When plots or rows are vacated by harvest, plant green manures or fall vegetables. Keep watering, and renew mulch that may have been beaten into the ground.

Field

Cut your second crop of hay when legumes are blossoming, which will be anywhere from four to eight weeks after the first crop, depending on what you're growing and how the weather's been. The procedure is the same as it was for the first crop, but the second cutting always seems to go a little easier than the first: the crop is lighter with smaller stems, so it is easier to handle and quicker to dry. You don't have the pressure to get it done before it gets overmature, either, since the second crop can "stretch" for a few weeks without becoming tough and unpalatable.

In parts of the Southwest with big irrigation systems, alfalfa may be cut as much as 10 times a year; in parts of the dry northern Plains, hay is traditionally a once-a-year, midsummer job. Dairy farmers in the eastern part of the country like to take four or more crops of alfalfa if they can. But for the small holder, two cuttings of hay in a season is generally plenty to handle. In a dry year when the pastures give out by late summer, the hayfields can be grazed instead of taking a second cutting — though use your judgment. If you didn't make enough hay on the first cutting to get you through winter, you'll have to buy more, and it will probably be scarce and the price high, since all of your neighbors are experiencing the same conditions you are. We often graze part of the hayfields in late summer and cut the rest for baling.

Bloat

Bloat can occur very quickly (within hours) in ruminants (grazing animals with four-chambered stomachs) when they are turned on to lush fine forage, most commonly young legumes. The swallowed forages form a mat over the contents of the rumen so that the animal can't belch out the gases that accumulate during digestion. When this happens, the rumen expands until it puts enough pressure on the lungs that the animal can't breathe and it dies. To prevent bloat, do the following:

- Grow mixed grass/legume hay; this is much less likely to cause bloat when grazed.
- Feed hay in the morning before turning livestock out on legumes.
- Don't graze very young stands of legumes; the older it is, the less likely to cause bloat.
- Do not graze legumes for 3 days after a frost.

Compost

Compost is a cornerstone of organic, sustainable soil building. Essentially, compost is made of organic wastes — such as animal manure, table scraps, and plant debris — put in a pile so soil-dwelling bacteria can transform it from a heterogeneous mess into a homogenous, dark-colored organic matter rich in nutrients. Compost may be applied at any time of year as a soil-structure builder and slow-release fertilizer. In orchards and hayfields, leave compost on the surface; in garden and fields, till it in for best effect.

The classic method for making compost is to build a pile a minimum of 3 cubic feet in size, in layers consisting of 6 inches of plant matter (such as old hay, leaves, coarse sawdust, and vegetable matter from the garden), 2 inches of animal manure (and bedding, if it's mixed in), then a sprinkling of soil and a sprinkling of mineral — either lime, wood ash, granite dust, or a similar amendment. Repeat the layers until the pile is at least 3 feet high. The soil contains the bacteria that will do the heavy work of decomposition. As this process gets under way, the pile will heat up enough to feel very warm to the touch just under the surface.

Turn the pile every few days up to every couple of weeks to speed decomposition and so that compost can be made several times in a season. Keep the pile moist, sprinkling it with water if needed.

People living on a small place often don't have the materials or the inclination needed to maximize compost production. Instead of building piles in a single day, many of us throw on materials as they are accumulated — in a bucket in the kitchen, inedible plant parts from the garden, manure from cleaning up the barnyard. Then we throw on some soil and wood ashes on occasion, and toss in some dry matter such as old hay or dry leaves to increase the carbon content, so the final result will retain nutrients better and the pile will rot down on its own without much further attention. This approach takes a lot longer, and the result may not be so evenly textured, but the soil and plants won't care.

You can build free-standing compost piles or put the compost in bins made of anything from wood pallets to chicken wire and old truck tires. Three bins is the standard: one to turn a pile into, another for the pile currently under construction, and the third for the pile that's cooking and will be turned soon into the empty bin.

If you observe the Final Rule of the National Organic Program, then you may apply compost made according to its specifications at any time. If you're more casual about your composting, simply follow the rule's specifications for applying manure and you will still be in compliance. The NOP Final Rule states that raw manure must be incorporated into the soil at least 120 days before the harvest of any crop intended for human consumption, or applied at least 90 days before harvest where the edible portion of the crop will have no contact with the soil, as with tree fruit.

Like many livestock owners, we also graze the hayfields early in spring to give the pastures a jump and set the hay back a little; this spreads manure on those fields at a good time of year. We'll graze the hay in the fall as well if the crop has grown back well by then. Whenever livestock is turned on pure stands of young legumes, there is a danger of bloat (for details, see Bloat, page 121).

Grain Fields

Fields where the small grains were harvested should be tilled and replanted, either with a forage crop or a winter cover crop. If there were significant weeds in the small grains, this is an opportunity to employ the "stale bed" method of weed control. To do so, cultivate the field repeatedly at intervals of a week or two to turn up weed seeds and get them to germinate, then kill the seedlings with the next cultivation.

Pasture

As in the previous season, continue to adjust the grazing rotation to grass growth. Cool-season forages will grow more quickly when the weather cools, but warm-season grasses will slow down. Stay on top of the weeds — they're still trying to set some seed before winter, and they will give up height in order to quicken the cycle.

Orchard

Mow the orchard ahead of harvesting early fruits to make it easy to move people and equipment in and out. This also makes it easy to spot the fruit you've dropped. The early-season apples are certainly ripening; since these types aren't generally good keepers in cold storage or good for making cider, eat some fresh and quickly put the rest into jars.

A Word about Unaltered Males

Breeding season for cattle is well under way and is approaching for sheep and goats, so it's a good time to review safety procedures around unaltered male animals — bulls, rams, bucks, boars, and roosters.

No matter how gentle your male, never turn your back — you can't ever trust him not to attack. Each year farmers get killed by bulls. Boars can kill you, too, and though rams and bucks won't, as a rule, they can put you in the hospital. Roosters can come after you with their beaks and spurs. Most roosters aren't too bad, but if you get a particularly nasty one, make it into soup and find another with a more docile temperament.

If you have a bull, ram, or buck out on pasture with the females, make sure there's a fence line nearby that you can roll under or an ATV on hand to hop into if you get charged. It's also a good idea to keep females between you and him, if possible. A good dog that knows its business can stop aggressive behavior.

Beeyard

Finish the first honey harvest, leaving a full super for each hive. Put on empty supers in anticipation of late-summer and early-fall blooms, as indicated by hive strength and activity.

Barn

In anticipation of fall vaccinations and weaning, assemble or tune up livestock handling facilities. Make sure ropes and latches are in good condition, gates and squeezes are oiled and working smoothly, and all the components — holding areas, crowding tubs, chutes, headgates — are aligned with each other.

If animals are not already accustomed to moving through your setup, start training them now. This is especially important with cattle, who are too big to push around. Cattle upset by unfamiliar things happening in unfamiliar surroundings can turn what should have been a quick and simple process into a rodeo, and you won't be able to stop them.

You can easily train animals by putting a daily grain ration or the water tank inside one of the holding pens and having the animals exit through the chute and gate. When they come up from pasture for a drink or a snack, close the gate on the pen behind them and leave the headgate open. After walking through every day for a few weeks, it will be much less alarming for everyone when you catch them in the chute or gate for a few moments to vaccinate or ear tag.

Choose a balmy, sunny day for harvesting honey, when the bees will be busy and happy and least inclined to bother you.

Handling Facilities

If you have just a few animals that are very tame, especially if they're halter-trained, you may not need any handling facilities. For larger flocks and herds, both humans and animals will be safer and calmer in a well-planned and well-built facility. At a minimum, this should include a holding pen, a single-file chute, and a headgate where you can catch animals as they come through and hold them securely. I've seen temporary sheep facilities made mostly of plywood that worked beautifully; for cattle, you need heavy wood or metal. I once toured a buffalo facility built out of used highway guardrails; not even those massive animals could budge that.

Temple Grandin has revolutionized animal handling in this country and around the world through applying her study and deep understanding of animal behavior to the construction of humane handling facilities. Grandin-designed facilities work with livestock's natural behaviors, such as the preference to move around curves and toward light, to create a calm and smooth handling process. Check out her book *Humane Livestock Handling* or go to her website (see Resources) for information on animal behavior and facility design.

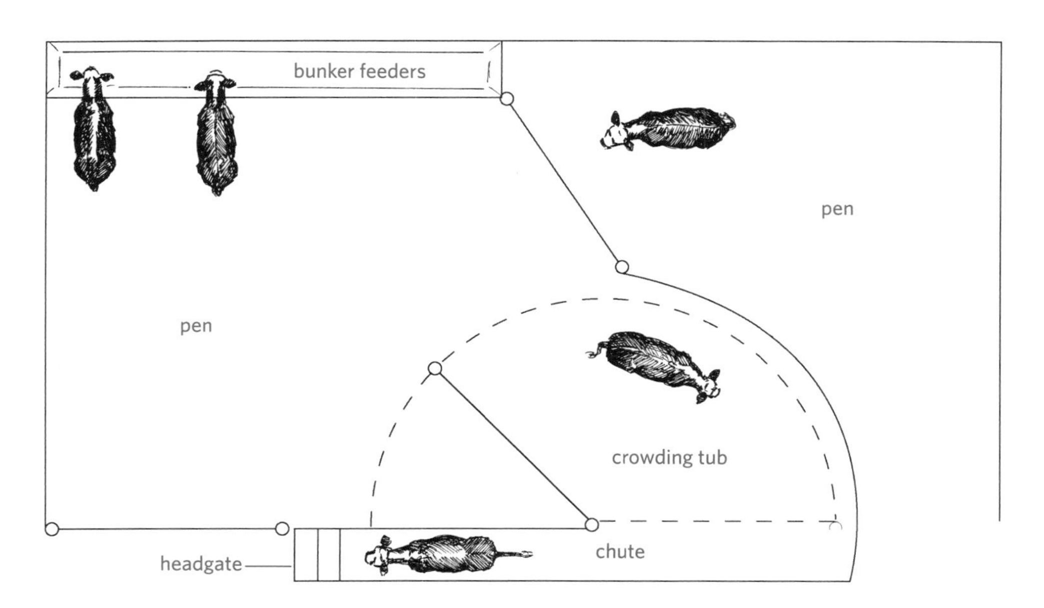

You can build an efficient handling facility for small herds and flocks in a relatively small area.

Coop

Move meat chickens to pasture pens as soon as they're nearly feathered out. Older laying hens are going into their annual molt; they'll quit laying eggs and be a little cranky for four to six weeks.

Equipment Shed

The summer routine of regularly checking oil and coolant levels, greasing frequently, sharpening mower blades, and keeping up with small repairs continues. When the final cutting of hay is done, those implements can be cleaned up and stored for the season.

Wetland Crossings

Late summer is often the driest part of the year, making it a perfect time to build and clear drainage ditches and install culverts. The lack of water makes it less likely that equipment will get stuck in the mud, and you can see what you're doing when holes and ditches aren't full of water.

Culverts are commonly installed in earth berms built to provide a road for equipment across a low wet spot, allowing water to drain from one side to the other while traffic above stays dry. This can fix problem spots in woods trails, cattle lanes, and farm roads. For woods trails that won't be used for logging equipment or other heavy vehicles, it can be much simpler to make wet spots passable year-round. If only foot traffic is expected and the bottom is firm, add stepping-stones. A load of fist-sized stones lets water flow through and allows for light vehicle traffic. An old favorite for swampy spots, where the bottom is not firm, is to make a corduroy road by laying poles (such as the trunks of straight young trees that you've been thinning) closely together across the mucky spot. (*Note:* In some areas, a permit may be required to build a wetland crossing.)

The most effective, and expensive, way to prevent equipment from getting stuck in the mud is to build a bank and culvert. Check with your Extension agent or local forester first to see if you need a permit.

Woodlot

If you have some cool, overcast days at this time, you can mow woods trails and even cut some firewood. This is especially important if your winters are wet and the ground isn't frozen, making woods work difficult then.

If it's a dry season, this is the best time of year to upgrade wetland crossings on your woods trails, dig ponds, and restore wetlands.

Wildlife Habitat

Water is a magnet for all types of wildlife, and digging a pond or restoring a drained wetland can increase the population on your land very quickly. Wetlands are federally regulated, and in many areas ponds are as well, so consult your state Department of Natural Resources and the U.S. Fish and Wildlife Service before beginning a project. Cost-sharing funds are often available, along with technical advice for siting a pond where it will hold water and receive enough drainage from the surrounding area to stay full most of the time. Technicians from these agencies will also calculate how big an area a restored wetland will occupy, and other details.

Except for the very smallest ponds and wetlands, the digging is done with heavy equipment. You can plant the raw soil on the banks with native species, or you can take your chances and see what grows on its own, though that can be an invitation to invasives like reed canary grass to move in.

Ponds intended to hold fish and amphibians should be dug deep enough that they won't freeze to the bottom where animals hibernate in winter. Ponds intended primarily for waterfowl can be shallow (turtles and large fish will eat young waterfowl). An erratic shoreline will shelter more animals; a smooth one leaves less habitat. An island or peninsula is attractive to nesting waterfowl.

To restore a wetland, you must block or destroy any drainage (surface or subsurface) that had been installed previously to drain the area. Restoring a wetland may involve giving up quite a bit of useful field space.

Another late-summer habitat-building task is to mow the food plots for wildlife that were planted last spring, in order to freshen them up for fall. If you didn't plant any food plots and would like to provide some late fall and winter feed for wildlife, you can till up plots and plant them now.

Topic of the Season

Tools and Equipment

Tools and equipment are central to working efficiently on the small holding. Certainly you could do all of the work by hand with the odd sticks and stones, but having implements and engines designed for the job means more work in less time with less muscle. I've shoveled enough manure by hand in my life that I appreciate a good shovel, but these days I would rather rent or borrow a skid steer to clean out the cow shed and barnyard in spring.

As we've become more mechanized over the years, we've become particular about keeping up on engine and implement maintenance and having a good tool set for both the routine tasks and repairs. But the first step in the process was finding equipment that fit our farm size and enterprises and that we could afford.

Although small-scale, especially walk-behind, equipment built specifically for small acreages is becoming increasingly available, it can be expensive. And if you're not close to a dealer, you may have to wait for parts when you need them. When it comes to doing the job well with the least amount of upkeep and storage space required, however, this class of equipment is perfect. Our acreage is a little on the large side for the small-scale stuff — we cut 30 to 40 acres of hay at a time, for

Most small farmers acquire a fair amount of tools and machinery over the years. Our old tractor handles less-demanding field work, while the ATV is a great labor-saver for livestock, fencing, and woodlot work. Tillers, trimmers, and mowers handle garden, lawn, and trails.

example — and on the small side for modern large-scale equipment. So we've relied on used, even antique tractors and implements that were originally built for farms of about our size.

We have found equipment at used equipment dealers and farm auctions, through want ads, and even sitting by the road with a For Sale sign on it. These days the Internet is a primary source, though I would still start with a dealer if you have one in the area, not only for their knowledge but also to have a relationship with someone who has parts and mechanics. You can often get a good deal at an auction, but buyer beware! Auctions offer no guarantees and no information on the condition of an implement. If you aren't qualified to judge for yourself, take someone with you who can assess whether something is functional or has problems. If there's an engine involved, start it and listen to how smoothly it runs. If you're in an area where there aren't dealers or many auctions, as is increasingly common now, then the Internet is the place to search. Just don't buy anything sight unseen!

Tools

With tools, you usually get what you pay for, so don't buy cheap. On the other hand, most of us don't need professional quality. Midrange tools from a well-regarded manufacturer

Every Year Is Different

Our plans and proposed schedules are based on average seasons: average date of last and first frosts, average coldest winter temperatures, average hottest summer temperatures, and average rainfall in different months. But anyone who has spent a lot of time outdoors for a few years knows from experience that an average season is a rare thing. Seasons vary so much in their length, abruptness, rainfall, and temperature that we become accustomed to dealing with the unusual.

Using a sustainable approach to farming and forestry gives you an advantage, because taking care of the soil and having a diversity of crops and animals build resilience into your system. Keeping soils in good heart and well covered prevents erosion in even the heaviest rainstorm. Having a high organic matter content and using mulch will conserve moisture in the soil in dry times, and applying livestock to the land in a rotational grazing system strengthens the sod and grows healthier animals.

But you can't always beat the weather. A severe drought may mean selling livestock until rain can revive your feed source; an abnormally cold winter can wipe out fruit trees even if they're normally hardy for your climate zone. A hailstorm can destroy large swathes of crops. Sometimes you just have to wait it out, clean it up, and replant and restock.

will do the job, though if you can't find exactly what you want in that range, I'd spend up rather than down.

Besides the basics — hammers, a selection of screwdrivers, pliers, and pry bars — it's good to have the following on a small to midsize farm:

- Grease gun with a flexible nozzle
- Set of box-end and open-end wrenches
- Set of socket wrenches, with both metric and American sockets, and spark plug sockets
- Adjustable (crescent) wrenches
- Set of Allen wrenches
- Bolt cutter and nut splitter
- Air gauge and portable compressed air tank
- Air compressor for filling the air tank
- Jacks and blocks
- Jumper cables
- Battery charger
- Oil filter wrenches
- Gapping tool for spark plugs
- Spring clip tool
- Impact wrench for bolts
- Grinder for sharpening blades
- A punch set for pushing out pins and aligning holes
- A set of pipe wrenches for turning round pipes and bars
- Automobile-type lug bolt wrench for tire bolts
- Bench vise for holding parts while they're sharpened or adjusted
- A small propane torch for heating frozen (stuck) bolts to loosen them
- Wire brushes for cleaning parts and bolt threads
- A shop magnet for finding stuff we've dropped
- Shop vacuum
- Power washer

Maintenance

The more maintenance you do, the fewer repairs you will have, and I hands-down prefer the first to the second. Maintenance begins with reading the owner's manual — or at least the sections that outline the maintenance schedule and list specs for oil, fuel, and parts. The next step is buying the supplies and parts you need on a regular basis: grease, engine oil, lubricating oil, and antifreeze and distilled water for radiators, as well as air and oil filters and the right size and shape of belts.

It's not hard to keep equipment running, but it does demand some regular attention to detail. If you can, always store it inside, where sun and rain won't cause brittleness and rust. Also keep equipment clean to extend its life, since accumulated gunk in the corners holds moisture and causes rust. We power wash the tractors but only blow off implements like the baler and grain drill, since they have too many nooks and crannies where moisture could sit.

Grease or oil moving parts as suggested in the manual, or more often on old equipment — this is probably the single most important thing you can do to keep implements running. Sharpen blades regularly. Check frequently to make sure bolts, belts, and chains are appropriately tight and do not show signs of wear. If they do, replace them.

Equally important to keeping an engine operational is to check the oil level frequently and regularly change the oil and the filter, if it has one. After this is done, check and clean or change the air filter and spark plugs as recommended in the manual, or at least annually. On larger engines with radiators and hydraulic systems, check the fluid level regularly, add as needed, and flush if necessary (though this should be a rare event).

Alas, even the best maintenance program can't prevent breakdowns. The minute something doesn't sound right or acts a little funny, stop! This will often prevent a small problem from becoming a bigger one. The next step is to trace the problem to its most likely source, which only gets easier through experience. At least try to narrow it down to the part or system involved — air intake and exhaust, engine lubrication and cooling, electrical, hydraulics, or power transmission. Much of the time you just have to poke around and look or listen for the particular part that's not working. Some of the time, especially with electrical problems, you have to work through a system piece by piece to find the bad connection.

And some of the time, you're better off getting help — from a neighbor, a friend, or a professional. Make sure you watch and ask a few questions, so that next time you can fix it yourself.

To keep equipment running well during periods of heavy use, you need to give it regular attention, such as scraping dry grass off the bottom of the mower deck.

Orchard

» **Mow the orchard again.**

» **Begin harvesting mid-season apples and grapes.**

Beeyard

» **Harvest honey** for the last time. Leave enough to feed the bees in winter.

Early Fall

Barn

» **Buy feed** if you're short.

» **Make sure you'll be able to move hay** to animals or animals to hay in winter bad weather.

» **Wean young stock.**

» **Castrate and vaccinate calves** if you haven't already done so.

Coop

» **Start collecting eggs again** as old hens come out of molt and new poults begin to lay.

» **Butcher meat chickens** that were started midsummer.

Equipment Shed

» **Do annual maintenance** on small engines.

» **Clean, lubricate, and store implements** as you finish using them for the year.

Woodlot

» **Cruise the woods** to plan where you'll cut firewood, harvest timber, and thin and prune in the coming seasons.

» **Mow trails one last time** if needed.

Wildlife Habitat

» **Plant winter food plots.**

Early Fall

Garden

The season is winding down. As vegetable patches or rows are finished for the season, remove what is left of the plants and apply manure, compost, or other recommended amendments before tilling and then planting winter cover crops or applying mulch for winter cover. Leave the hardy vegetables — kale, carrots, and similar crops — until mid-fall, after freezing temperatures stop growth (in warm regions many such plants can be overwintered, especially with a little mulch as insulation). Pumpkins and winter squash will keep better and are easier to harvest after a hard frost has killed the vines. Leave any flowering plants and vegetables until late fall — they'll help feed pollinators at a time of year when blooms can be scarce.

Field

Plant winter grains. Plant cover crops in fields being fallowed over the winter.

Rest hayfields for four to six weeks before the first expected hard frost, so that the plants can store reserves in their roots that will keep them alive through winter and give them a quick start in spring. We've found that a light grazing during part of this time on our mixed grass/legume hay stands does little harm, however, and allows the pastures rest time to regrow for stockpiling after the growing season ends. Grazing hayfields also puts manure where it's needed and frees up time for working in the garden and orchard.

Don't graze the hayfields too short — the legumes need that insulating blanket of foliage to protect them from freeze/thaw cycles that heave the soil and break roots, especially in heavier soils. And never turn out animals on legumes first thing in the morning, or after a frost, due to the danger of bloat (see page 121). Wait at least three days after a frost, and on all days make sure livestock fill up on hay before being turned out.

Pasture

If the weather cools off and gets a little wetter, cool-season forages will resume a fairly rapid growth for a few weeks (if fall is warm and dry, this won't happen). More pasture growth extends the grazing season. Or you may stockpile paddocks for late fall and early-winter grazing. On a small place, this is best accomplished by grazing the hayfields in late summer and early fall, letting the regular pastures rest and regrow as a stockpile for mid-fall to early-winter grazing.

Orchard

Keep the orchard mowed from the beginning of fruit harvest until the grass quits growing for the year. Peaches, apricots, plums, and cherries are done by now, and pears are winding down, but prime grape and apple picking season is under way, and most nuts are beginning. Among the berries, only cranberries are left to be gathered.

We have far too many apple trees for household needs but not enough to make it worth the effort to sell them — a common problem. We solved this by holding an annual apple party, which we schedule for just before or after the first expected hard frost, when the varieties we've planted should be at their peak. It's also usually some of the finest weather of the year. Early on party day, we haul out and clean the cider press and every bucket and basket we can find. As our relatives, friends, and neighbors arrive, we first have them pick a grocery bag or two of the very best apples to take home for fresh eating. Then we break for lunch, after which the serious work begins: moving apples from tree to cider press and apple butter pot. Because deer move through our orchard, apples on the ground are potentially contaminated by their manure, so none of these are used for human consumption — the cattle get them, along with any other fruit not good enough to go in the press or the pot. By the end of the afternoon, the herd is lined up along the pasture fence waiting for the kids to toss them more.

An apple party, where your guests take turns manning the cider press, is a fun way to make use of a large apple harvest.

The pomace from cider making and the skins and cores from apple butter making can be fed to livestock, composted, or left in the woods for local wildlife. If you have a game camera, focus it on the apple pile and see who visits.

If the weather holds, we end the day with a picnic dinner and a bonfire in the yard. If it rains, we move the eating and talking inside. Bees and wasps attend our apple party in large numbers, but they seem to be as happy as we are, and in all these years, we've only had two stings.

Beeyard

The final honey harvest should take place a month before the first expected hard frost, leaving enough filled or partially filled supers to feed the hive through winter. How much exactly depends on the length and severity of your winters, so consult with state or local beekeeping associations. Hives without enough honey should be fed heavy syrup. Check hives for mites and medicate (mix the medicine with the syrup) if mites are present.

Barn

Early fall is a traditional time to wean calves, kids, and lambs so that they have a few weeks to recover before the weather gets chilly and the extras are shipped to market. Beef cattle calves are often also vaccinated and the bull calves castrated at this time, if they weren't at birth. Heifer calves must be given the brucellosis vaccine between 4 and 12 months of age for it to be effective. This must be administered by a veterinarian, who attaches a metal ear tag to the animals as proof. (Sheep and goats most often are given their annual vaccinations 2 to 4 weeks ahead of giving birth, and the pigs don't need vaccinations, since they'll be butchered soon.)

Many of the annual livestock vaccinations require booster shots 2 to 12 weeks after the initial shot, depending on the vaccination. After that, a single yearly shot is all that's needed for most vaccines. It's convenient to separate mothers and youngsters at booster vaccination time, since they're being handled anyway. This is also an excellent opportunity to apply ear tags, deworm, and trim hooves if those items are on your to-do list. Beef calves can be castrated at the time of the initial vaccinations, returned to their mothers to help them recover from the stress, then weaned when the booster shots are given. If you plan to keep your young cattle through winter, you can let their mothers wean them — most will do this naturally in late winter, a couple of months before calving again.

When working livestock, the most important thing to remember is to stay calm. Move smoothly and slowly, and don't yell and start slapping butts. If your animals are accustomed to being tied or to walking through a chute and headgate, the process will go smoothly. If you have an animal that just won't accept being handled or herded, cull it. A replacement is cheaper than a trip to the emergency room after being kicked, butted, or run over.

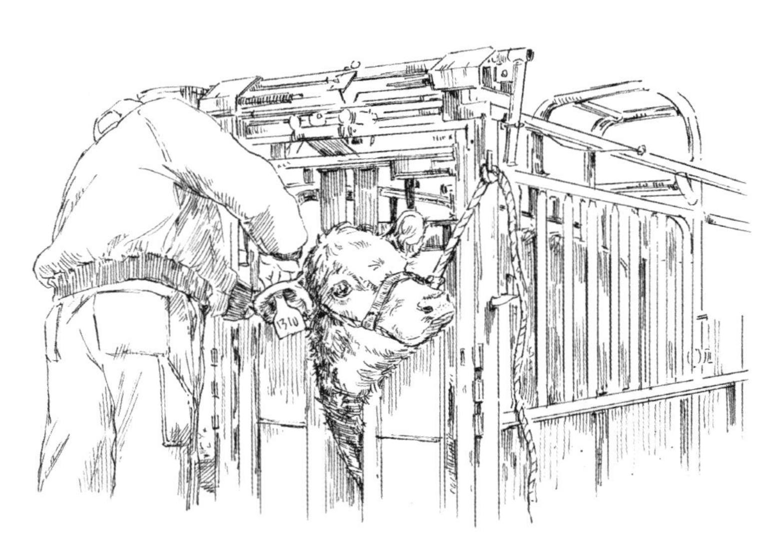

A chute and headgate make the fall chores of vaccinating, castrating, eartagging, and deworming a relatively smooth process.

Weaning Techniques

There are two good ways to wean animals. If you have a small place, put mothers on one side of a fence and babies on the other. You need tight fences, and an electric wire helps, too. There will be quite a bit of bawling for a couple of days (this is not the time to have overnight guests), but since they can all see each other and lie down close together, the animals will be less stressed and will calm down after about 3 days.

The second, more traditional, method of weaning is to completely separate the mothers and babies, either by trucking the babies to a distant location or locking them in a pen or barn away from their mothers. On a small place, they'll still be able to hear each other, so the bawling can go on for quite a few days.

Coop

If you started a new batch of meat chickens in midsummer, they're about ready to butcher by now. I prefer the moderately cool mornings of early fall for this job, when there are few flies around and it's not so cold that my hands are chilled by all of the water involved. For a discussion of how to butcher chickens, see page 110.

Molting hens can be a little grumpy, so be kind.

This year's poults should be starting to lay, and older hens should be coming out of their annual molt and starting to lay again.

Equipment Shed

As the growing season winds down, order or pick up the parts and supplies you'll need for doing annual maintenance on all equipment, readying it for winter storage. This includes air and oil filters as well as all the various weights of engine oil necessary for different engines — such as mowers, trimmer, tractor, tiller, utility vehicle, and chain saws. As you finish with a piece of equipment for the season, put it in the queue for cleanup, maintenance, and winter storage. Now is the time to change oil and filters, service the air filter, replace worn belts and chains, check radiator and hydraulic fluid levels, check air pressure in tires, and do general cleaning and tightening up. We have quite a few small engines that get heavy use during the warm seasons, not to mention the tractors, haying equipment, and some tillage implements, so we like to start this chore early in order to finish it before colder days make it unpleasant to do bare-handed work. (For more on equipment maintenance, see page 128.)

When parking implements for winter, it helps to put in last the implements you'll want first in spring, to avoid the hassle of having to pull out equipment to get at what is needed in the back.

Early Fall

Woodlot

If it's dry, this is a good opportunity to improve trails or create new ones, since you still have time to seed the raw soil so it can have a cover before winter. If your trails cross wet spots and you didn't improve the crossings in late summer, this is still a good time to lay down rock, corduroy (made from wood poles laid tightly together), or a bank and culvert (see page 126). And while you have the equipment and operator out there, dig any planned ponds.

Prepare for upcoming woods work by marking trees for firewood and timber harvests and targeting the stands you want to thin and prune in the coming season. Lastly, the job of controlling invasives never stops — look for any problems and be ready to mow, pull, or poison. We don't use poisons, but many woodland owners faced with overwhelming populations of invasive plants find this is the only practical solution. This is the perfect time of year to poison woody invasives by lopping off their tops and dabbing the chemical on the cut stump. The plant is busy recovering nutrients from the leaves to store in the roots for winter, and the poison will go rapidly to the roots. All you need is a pair of loppers for the cutting, a sponge tied to a stick, and a small bucket of chemicals that can hang from your belt.

Wildlife Habitat

Till plots and plant winter food for wildlife, such as winter grains and hardy legumes for grazing.

Weed, Pest, and Disease Control

On a diversified small holding where the owner is proactive about preventing weed, pest, and disease problems, there should be few issues, but not none. Sooner or later, a bug, weed, or illness will likely show up that has to be dealt with. Sustainable farming advisors consider integrated pest management, or IPM, to be the best management technique both for preventing and solving problems before they get out of hand, with a minimum of chemical inputs. Organic IPM puts more emphasis on prevention, since treatments are limited by the prohibition against synthetic chemicals except under very specific circumstances.

Birds can provide significant bug control; all you have to do is provide the right habitat.

An IPM program has six components: planning, setting action thresholds, monitoring and detection, proper identification, action, and evaluation of results. On a small place, the program does not have to be formal or even particularly detailed until you're facing a definite problem.

Planning

Planning is two-pronged: (1) maintaining healthy soil and raising both plants and animals in a way that minimizes the potential for problems, and (2) being prepared for problems that may occur.

Maintaining soil health and fertility (as covered in chapter 1) is based on a continual program of incorporating organic matter and nutrients into the soil and keeping it covered so it's protected from erosion. There are various aspects of raising plants to minimize problems such as weeds, insects, and diseases. A plan should include crop rotation, selecting varieties for resistance to diseases, proper timing and spacing when planting, utilizing trap crops and beneficial insect habitat, and good sanitation (removing and disposing of potentially diseased plants, fruits, and prunings).

A plan for raising healthy animals (discussed in chapters 3 and 4) should include selecting breeds adapted to your region and providing free access to pasture, shelter from inclement weather, adequate and appropriate diet, clean living quarters and water, appropriate salt and mineral supplements, and vaccinations.

To anticipate and prepare for likely potential problems, research what the common problems are in your area and what controls are available for dealing with them. Ask your veterinarian what diseases and parasites are the biggest headaches in your neighborhood. Plant problems are a constant topic for gardening associations, fruit growers organizations, and Extension agents specializing in crops and horticulture. The Internet is full of information, too, but be sure to use reliable sources that are specific to your region — such as local, state, and regional Cooperative Extension Services, Master Gardener associations, fruit growers associations, and sustainable farming organizations.

If a problem does crop up, your options are cultural, biological, and physical controls. Chemical controls are a last resort.

CULTURAL CONTROLS include establishing habitat for beneficial insects, planting trap crops, and removing species that serve as hosts for diseases that afflict your crops (avoid growing blackberries near white pine, for instance, since the first is a host for white pine blister rust). For animals, this category includes changing the rotation schedule so animals aren't in the pasture when parasite larva are ready to be ingested.

BIOLOGICAL CONTROLS include releasing beneficial insects; applying sprays of fungi, bacteria, and viruses that afflict specific pests; and using pheromone traps.

PHYSICAL CONTROLS include using row covers to prevent insects from landing, spraying kaolin clay compounds on fruit for the same purpose, screening around fruit tree trunks to ward off rodents, putting up traps like saucers of beer for slugs, building fences to keep animals out of the garden and orchard, mulching and mowing to suppress weeds, and using guard animals to keep predators away from livestock. These types of controls also include things like handpicking and vacuuming bugs and mowing, cultivating, and flaming weeds.

Setting Action Thresholds

The second step in an IPM program is to decide how much damage you're willing to put up with before you do something about it. On a large operation, this is a mathematical decision based on how many insects you are collecting in monitoring traps or the percentage of a crop that's afflicted with a particular bug or disease. On a small place, it's more likely to be a seat-of-the-pants decision to do something before a problem gets any worse. If you had a problem last year that got out of hand, it's a good idea to anticipate its happening again this year and be ready for it early.

Monitoring and Detection

On a small place, monitoring and detection translates to "taking a walk around your land every day and checking things over." To recognize that something is wrong, you must have a detailed knowledge of what's normal, and this is only developed through experience — much of which can be gained by taking some time every day to just look. Problems can also be fast moving, making daily monitoring very important. An insect problem can blow up almost overnight if you don't catch the appearance of the egg cases or tiny larvae.

Proper Identification

Put a bug in a jar, or pick a sick plant, and compare it to identification guides. If there's any doubt, involve an expert, such as an Extension agent, forester, or veterinarian.

Action and Evaluation

After identifying the problem and deciding to do something about it, the next step is to do it. Lastly, after the situation is over, jot down some notes about what you did, when, and how well it worked. This will help you do even better next season.

Pheromone traps function as early-warning systems for insect pest problems.

Weeds

Weeds are the most obvious and persistent pests in gardens and fields, sucking up nutrients you want for your crops and crowding access to sunlight, water, and air. Options for controlling weeds in addition to cultivation (as discussed on page 142) include:

- **The "stale bed" method** involves tilling bare ground repeatedly to bring weed seeds to the surface so they sprout, then killing them with cultivation. Do this for several weeks before or after growing a crop.
- **Plant and till under** a weed-suppressing crop such as buckwheat or oats several times during a growing season.
- **Space crop plants** so that as they mature, they will shade out weeds.
- **Plant vegetables earlier or later** than pests are expected to be active.

CHAPTER TEN

Mid-Fall

Begins with the first hard frost, which kills the corn. Deciduous trees shed leaves; average daily air and soil temperatures trend to below 60°F/16°C; cool-season grass growth slows and stops.

THE FIRST HARD FROST signals the start of mid-fall and the end of the growing season for all but the hardiest plants. Apple picking continues, but the pumpkins and winter squash are left in the sun for a couple of weeks to cure for winter storage. As the leaves on the hardwoods turn color and drop off, wild nuts are gathered and hunting for small game begins. Wear a blaze orange or green cap, vest, or windbreaker if you're out in the woods, so hunters can spot you.

This is the season to store equipment and fill the cold cellar. Drain and store garden hoses, too. Cut and stack firewood, and prepare winter quarters for livestock. The harder your winters, the more you will appreciate being prepared for the ice, snow, and frozen ground.

Mid-Fall

FOCUS ON:
orchard • barn

Garden

» **Harvest storage crops** and move them inside.

» **Finish cleaning up** garden debris.

» **Spread compost and manure.**

» **Till** if you like.

» **Cover soil with mulch** if you didn't plant cover crops.

Field

» **Harvest begins for grain corn;** silage corn is already done.

Pasture

» **Take down temporary electric fencing** for the season as livestock graze each paddock for the final time.

Orchard

» **Finish the apple and grape harvests.**

» **Put the orchard and vineyard to bed** for winter.

Beeyard

» **Feed syrup,** if desired, to ensure a good honey supply for the hive through winter.

» **Install entrance reducers** to help keep cold out.

Barn

» **Finish preparing winter quarters** and feeding areas, making sure you'll be able to move hay to animals or animals to hay in bad weather.

» **Put heaters in water tanks.**

» **Ship any animals you don't plan to butcher** or carry through winter.

» **Butcher or ship pigs now,** or next season when it's cooler and they're heavier.

Coop

» **Clean the coop** and put down fresh bedding for winter.

» **Put a heater under the water fount,** and hang heat lamps if desired.

» **Move pasture pens out of the way.**

Equipment Shed

» **Finish cleaning,** doing annual maintenance, and repairing all equipment (wood splitter, chain saws, snowblower).

» **Store equipment for the season** or put it somewhere handy if it will be used during winter.

Woodlot

» **Pull woody invasive plants** if there's time and the soil is soft.

» **Start woods projects** and make firewood if time allows.

Wildlife Habitat

» **Build brush piles** with debris from woods projects.

» **Build sheltered winter feeders** for ground-feeding birds.

Garden

The first hard frost signals the end of the gardening season. Only hardy, late vegetables — kale, parsley, some of the root crops — are still green. Leave pumpkins and winter squash in the sun for a couple of weeks after the vines die back; this will cure them so that they will keep longer in storage. Also leave the late-blooming flowers and vegetables (broccoli flowers are very popular) for the occasional pollinator — bumblebees, especially, are active until very late in the season.

Finish clearing off the season's debris to break up pest life cycles, then apply compost, manure, and any other amendments indicated by the soil test. Till, if you like, to destroy eggs, larvae, or adults of those pest species that overwinter in the soil. If you don't have pest and disease problems, and you do have toads and salamanders overwintering in your soil, leave the tillage until spring. Those critters will repay your care in insect consumption. Lastly, mulch soil that isn't already protected by a planted winter cover crop.

Extending the Season: Fall and Winter

You can keep many vegetables growing into winter inside cold frames or similar structures. The frames should be sturdy enough to withstand substantial snow and/or ice loads. You can put many types of herbs in pots and grow them through winter on a sunny windowsill (chives and parsley are favorites).

A pumpkin patch lights up the fall garden.

If it's too late to plant winter cover crops, put down a light to heavy mulch of straw, old hay, or leaves to preserve your soil resource and add organic matter when you work it into the soil next spring. If you want to have very early greens in spring, place a heavy mulch over a small bed to keep it from freezing — you'll be able to plant there when the days begin to warm in late winter, before the rest of the garden soil has thawed.

Field

Harvest field corn (usually grown to feed livestock) between the time the plants are killed by hard frost and when the weather becomes too wet or the snow too deep to move equipment through. Fields are harvested by combines, which detach the kernels from the plant and the cob; by mechanical corn pickers, which harvest the cob but don't strip the kernels; or by hand, which leaves both kernels and husk on the cob. Traditionally, handpicked corn was piled and husked through winter. You can store husked corn in a roofed, open-air crib and feed it to livestock on the cob, or you can run it through a mechanical corn sheller (if you can find one) to strip the grain from the cob.

You may bale the dry corn stalks for animal bedding, or chop them and leave them in the field as winter cover, which will become organic matter when tilled under in spring. Don't leave crop residue in the field if it's providing habitat for pests or diseases that could affect the next crop in the rotation, or if you will be growing the same crop on an adjoining field next year.

A small patch of field corn can be harvested by hand.

Green Corn

Green corn is frequently harvested for silage in livestock areas, which is why you'll see equipment in the cornfields in late summer and early fall. Silage is made by chopping the entire plant, then blowing it into a silo, bunker, or giant plastic bag. There, the high moisture level and lack of air (silage packs tightly) cause it to ferment. The distinctive smell means green corn has turned into a highly palatable and nutritious animal feed.

If you till under the crop residue this fall, which will save time next spring as well as help to control soilborne insects or diseases, then plant a winter cover crop. An annual crop will die over the winter and, unless it's very heavy, you may be able to plant directly into it next spring, without additional tillage. A perennial crop will have to be tilled under.

Pasture

With the grass-growing season nearing its end, grazing livestock are finishing their final rotation through pasture paddocks. Take down and store the temporary electric fence as they finish a paddock to make your fencing materials last longer, and to make it quicker (fewer gates and corners to mess with) for you to do any final mowing of pasture weeds or overgrown grasses. If you have a lot of legumes (especially alfalfa) in your pastures and hayfields, don't graze or mow too short. An insulating layer of several inches of stubble protects against frost heaving in late fall and early spring, which injures and kills these plants.

Livestock that have been grazing hayfields, harvested cornfields, or other plantings (some winter grains and other cover crops will tolerate grazing at this time) can return to the pastures as they finish other fields and be rotated through stockpiled paddocks until those are used up.

This is also an excellent time of year to clear overgrown vegetation off permanent fences with mower and clippers (I always find myself pulling grapevines off the wires about this time). This will greatly increase the life span of your permanent fences.

Make hay available now in the barnyard or other area used daily by livestock, so they can gradually transition from green pasture to dry hay. This keeps their intake even and prevents digestive upset.

Orchard

Once you have picked all of the apples, bottled the cider, and put all of the butters and jellies and jams in the jar, get the orchard ready for winter. Clean up and dispose of any remaining fruit to interrupt pest and disease cycles. If the grass has grown, mow a final time to discourage rodents and to shred leaves to limit overwintering of apple scab spores (you could rake, too, if you only have a few trees or access to a mechanical rake). Paint trunks with white interior latex paint mixed 50-50 with water for sunscald protection. Check the screening around the base of young tree trunks to make sure it's tucked firmly into the soil and there are no open seams. Then (if you haven't already) put 6-foot-tall hoops of woven wire fencing around young trees to protect them from deer, or install 3-D deer fencing or a permanent fence around the entire orchard.

Beeyard

If hives might be short of honey for winter, feed them syrup. Remove mite medications. If you like, install an entrance reducer to help keep the hive warm and discourage rodents. Don't open hives unless absolutely necessary, since the bees have been sealing the cracks with propolis to protect against winter drafts. In cold climates, many beekeepers wrap their hives with an insulating material as the weather gets colder.

Barn

For March lambs or kids, breed ewes and does in October. For April lambs and kids, breed in November. Goats are often taken to the buck and may even be boarded at his home until bred; with sheep, the shepherd more often brings the ram to the ewes. You can use marking harnesses on rams to track which ewes have been bred. Change the color every 16 days.

Finished (fattened) meat animals are traditionally shipped to market or butchered in mid to late fall, just as the pasture runs out and more expensive stored feed (in the form of hay) has to be fed. This is still an excellent idea for keeping winter chores and feed expenses to a minimum. For home butchering, the weather should be cool enough that hanging meat won't spoil; air temperatures in the 30s and low 40s (–1 to 7°C) are ideal. If you don't do the job yourself, make an appointment well ahead of time with your processor to pick up the animal(s) before gun deer season starts. That's the busiest time of year for any local meat plant. If you wait until then, it may be weeks or even months before the processor can pick up the animals and you can pick up your meat.

Prepare winter shelter and feeding areas well before it gets really cold and miserable. Repair and tighten pens, sheds, and coops. Pigs, poultry, and goats especially need draft-free, well-bedded housing. Where temperatures will remain below freezing for long periods, insulate waterers. A heater in the tank keeps liquid water available at all times, though you'll still have to fill it with a hose every day. If you want to keep your electric bill down, you can just haul water twice a day.

Cattle and sheep can be outwintered (fed outside through winter; see next page) as long as they can get out of the wind and stay dry. In areas where there is rain or ice rain in

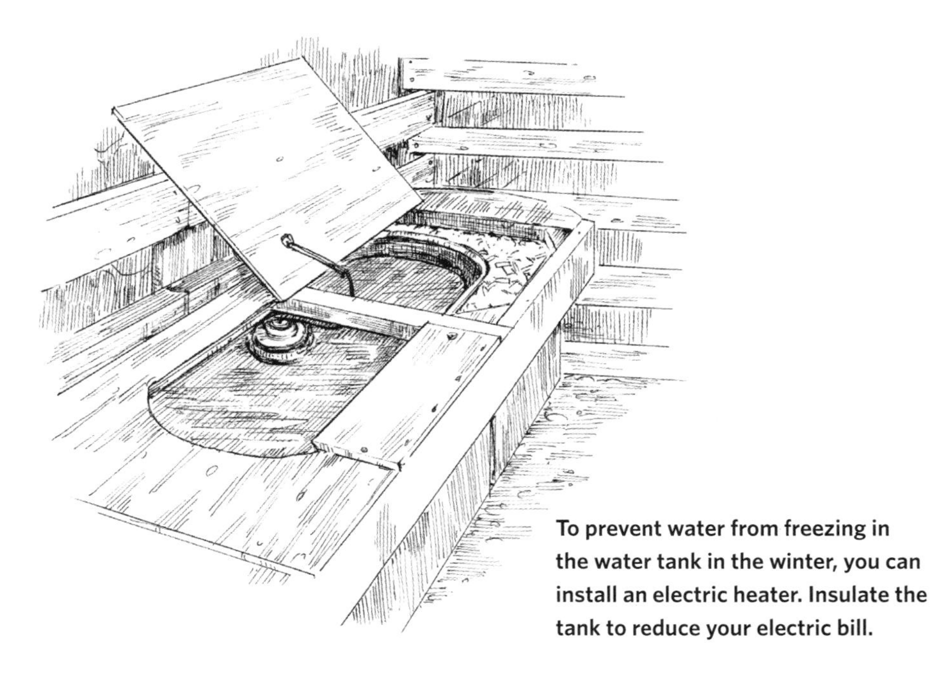

To prevent water from freezing in the water tank in the winter, you can install an electric heater. Insulate the tank to reduce your electric bill.

winter, this means having a shed available for their use. You can haul feed to the animals' designated wintering area as needed through winter, or you can set out round bales ahead of time, put an electric fence around them, and move the fence and hay feeders as needed. The latter approach means you don't have to start a tractor all winter, and the manure will be spread fairly evenly across the area.

We outwinter cattle on the poorer areas in the fields and pasture, which puts the manure where it's needed most. We've also found that the cattle stay cleaner and healthier than they would in a barn. The poor hay from the

Tips for Outwintering

- Set up a feeding area in the hayfields, if possible — the pastures are getting plenty of manure already.
- Choose a well-drained area, preferably with low fertility.
- Set out enough bales to get you almost through an average winter. Make an educated guess, since the actual rate of consumption will depend on how severe the winter is, whether you're feeding young stock or mature animals, and how much they're growing through winter. Have some feed in reserve in case of a very severe winter or a late spring. (I calculate five to six round bales per cow per winter, if the round bales are 5 by 5 feet and I feed hay from late October to mid-April.)
- Where ground is frozen, you can space bales more closely and have higher stock densities (I place bales 15 to 20 feet apart). If ground does not stay frozen reliably through winter and you want to avoid replanting in spring, space the bales more widely and limit time in any one paddock to 2 to 3 weeks.
- Fence the bales on three sides (where the ground will be frozen you must put up a ground wire if using electric fence).
- Run a temporary wire on the fourth side, to be moved back a row at a time as you feed the bales.

Placing bales for winter feeding before winter means the livestock move to the feed, instead of you bringing it to the livestock. You simply move the electric fence and the feeders by hand instead of starting a tractor, and there's no manure to spread in the spring since the livestock have spread it for you.

bottom and outside of the bales that gets left by the cattle serves as fresh bedding every time I move the feeders, and the animals are out in the sun and not in a moist, dim barn. In spring we disc the manure and old hay into the soil and reseed the area.

Outwintered livestock still need shelter from wind and wet, such as a simple three-sided shed or even a windbreak fence (tall and porous) in an L-shape that allows the animals to get away from the wind no matter what direction it's coming from. The open side of the shed should face away from the prevailing direction of storm winds, which is probably not the same as the prevailing direction of fair winds. And don't face it north, where the coldest winds come from. The shelter does not have to be immediately adjacent to the feeding area; it just needs to be close enough that animals will head for it in bad weather. I've found that beef cows don't mind a quarter-mile walk to their shed, but young dairy heifers need shelter to be much closer. Don't simply provide access to a woodlot; the animals will gradually kill the trees.

Coop

Clean out the coop and put down fresh bedding for winter. Put a heater under the waterer (available from poultry supply companies) or bring in buckets of water at least twice a day in freezing temperatures. A heat lamp isn't strictly necessary and will add to your electricity bill, but in return you'll get more eggs through winter and fewer frostbit combs.

Equipment Shed

Finish annual maintenance chores and store equipment. Any equipment that you will use during winter should be next to the door. Always park the tractor where you can get at it with another engine and jumper cables if it doesn't start in spring. Make notes of what parts and supplies you should order for next year.

Woodlot

This is the most pleasant time of year to be in the woods, no matter what you're doing! Gather wild nuts for eating or planting. If you're a hunter, this is prime season for small game: as the leaves start to turn and thin out, you can see farther into the woods. (Be sure to check the regulations for opening dates, and obtain a hunting license.)

Leaf drop is also a good time to mark trees for cutting and to lay out new trails. Begin making firewood and thinning trees if you have time. If the ground is reasonably moist and the bugs have dwindled, continue to cut and pull invasive species. If it's dry and the roots are breaking off, leave them until spring.

Wildlife Habitat

Build brush piles for wildlife shelter as discussed on page 157. You can also build sheltered feeders for ground-feeding birds like grouse and turkey. These usually consist of a slanted roof that drops to the ground, with the open sides facing away from the prevailing winds.

Topic of the Season

Managing Your Woods

Woods enrich a rural acreage, providing firewood, game for the table, fence posts, and occasional timber for boards, not to mention beauty, privacy, and extras like berries, nuts, and wild herbs. Unfortunately the vast majority of privately owned woodlands in this country has suffered from generations of poor logging and grazing practices and general neglect. The good news is that a woods owner can turn things around with some knowledge of where to start and a commitment to regularly putting in some time among the trees.

In some ways, managing a woods is similar to tending a garden. You're dealing with a variety of plants — both trees and understory — that each need their share of sunlight, healthy soil, and water to thrive. Like vegetables, trees vary in their soil type preferences, ability to grow in shade, and tolerance for crowding. In a garden, weeds can smother the vegetables. In a woods, invasive species can crowd out all other growth, preventing new trees from growing to take the place of old ones. If this happens, the forest has no future.

A landowner's task in the woods is to encourage a diversity of trees and understory plants, reduce crowding, raise high-quality trees, and control invasive species. The reward is a resilient and productive woods that benefits both wildlife and the humans who are its stewards.

In a nutshell: Take the worst; leave the best.

Defining Healthy Trees

To leave the best trees and take the worst, you need to know how to tell the difference. A healthy tree is straight, with a full crown, and produces plenty of seed to feed wildlife and germinate more trees like itself. Trees can be unhealthy due to disease or infestation, some signs of which are fungus growth, discolored or weeping bark, and holes made by bugs. Trees can also be stunted from overcrowding and losing the race for sunlight. These trees are small for their age with old bark, and they do not grow noticeably for years and even decades.

Left to their own devices, forests have mechanisms for relieving overcrowding, such as wind, fire, pests, and diseases. These forces get rid of weak trees, but too often they also get rid of all of the trees. These are natural rejuvenation processes, but despite their long-term benefits, none of us really want to have any of them occur in the woods around our homes. A better alternative for the people who live near these woods is called sustainable forestry, a set of principles and methods directed toward mimicking, in a much tidier manner, the natural processes that keep a forest healthy and growing. Though individual trees get old and die, the forest survives.

A thin crown with lots of dead branches (left) or a trunk with growths or oozing (right) indicate an unhealthy tree.

Applying Sustainable Forestry on the Ground: Thinning and Pruning

To get your mind around the plan for a large woods, it's very helpful to have a map of the different stands of trees and a written plan for what you should do when in each area. If you have just a few acres of woods, it's not necessary to be this formal — as you get to know your woods and what can be done, you'll see what ought to be done. You can take your time in the woods — there aren't the daily chores and tight timetables you have with livestock and other crops.

The first step is to thin crowded trees, then prune the best young trees. Thinning makes firewood and pruning makes high-quality future timber for sale or home use.

Ideally, you start thinning a stand of young trees when they begin to show signs of stress from crowding, such as slow growth, with some starting to die or showing signs of stunting. How much to thin depends on the size of the tree, but a good rule of thumb is that there should be open space on at least three sides of the crown. Leave the strongest, biggest trees and remove those that are being shaded by their neighbors, as well as those that are crooked, cracked, leaning, or looking sickly. The top of each of the remaining trees should be in the sun. You can leave the trees where they fall to deter deer (though this can be a fire hazard); pile thinned trees too small for firewood in brush piles for wildlife; or stack them for burning when it's safe to do so — when snow is on the ground or the ground is wet from rain (for details, see page 157).

You can thin bigger trees as well, but felling them takes more expertise and attention to safety (see page 159). These, along with trees that have fallen on their own, are your primary source of firewood. Larger trees that have been down for a while and are beginning to rot make poor firewood but are great for many types of wildlife; leave those big trunks where they lie.

Every site and every species is a little different, so it helps to learn the quirks of different species to get the job done right. For example, don't thin aspen at all; those species take care of the job on their own. Thin oaks only when they're above deer browsing height (roughly 4 feet) and it's clear which are the strongest growers. Thin pine to regular spacing. We also thin for species diversity and to favor species like red oak that grow well on our light soils and have many wildlife benefits and a high timber value.

You prune to improve the quality of timber that a tree will one day produce. Branches create knots in the wood; the fewer the knots, the more valuable the boards will be. When a tree is still young (a trunk diameter of less than 5 or 6 inches at chest height), prune off all branches to a height of 9 feet — the standard length for a log being sawn into boards. Pruning is a big job, so concentrate on the highest-value species (ask a forester or logger which wood is most valued in your region). Pruning also reduces fire hazard by reducing burnable material that could allow a low-intensity ground fire to leap into treetops and get out of control.

Getting Rid of Debris: Rotting, Burning, and Brush Piles

Thinning and pruning create a lot of woody debris, which can be a fire hazard and an impediment to moving through your woods. There are three good ways to deal with this: let it rot, burn it, or pile it for wildlife shelter.

In a moist climate (generally east of the Mississippi and on the ocean side of the mountains in the West), wood gradually rots back into the soil, enriching it with organic matter and nutrients. Small-diameter softwood like pine disappears in a few years, while large-diameter hardwood may last a couple of decades or more. Those big logs on the ground are valuable to wildlife, especially amphibians, so it's usually best to leave them alone. The small stuff can be an excellent nursery for young trees, since hungry deer don't like to navigate the tangle of branches, but it's also a fire hazard and kind of ugly. Though it's labor intensive, we usually pile the small stuff or stack it near a trail for burning.

Burn only when the ground is snow covered or the entire woods is wet: too many forest fires are started by people burning debris during a dry period. Choose an open area with no overhanging branches.

A well-located brush pile can provide shelter for many species.

Brush piles provide shelter for many species — including red squirrels, rabbits, and badgers — and serve as good incubators of bugs, which provide a lot of food for species higher up the food chain. In general, follow these guidelines:

- Crisscross large-diameter wood (and any rocks you have around) at the base to provide spaces for shelter. If you don't have a lot of big wood, use crisscrossed bundles of small stuff as a substitute.
- Crisscross small-diameter wood on top for a protective thatch
- Make piles big and broad rather than high. Small piles can be death traps for little animals being pursued by predators. The pile should be big enough that creatures taking shelter can't be easily dug out, or, if they leave on the opposite side, it's too far away for a predator to leap over and catch them.
- Place piles near water, food sources, and/or other cover if you want them to get used.
- If you don't like looking at them, put piles where they aren't visible. If you do like watching for wildlife activity, place piles where you can see them from the trail.
- Place piles on slopes or higher areas, so they aren't flooded or damp.
- Add additional materials through the years as the piles settle and decay.

Invasive Plants

Removing invasive species of plants gives your forest a future. If these smothering, aggressive plants are not removed, there will be no new trees to replace those that have reached the end of their lives.

Each region has its own cast of criminal plants, and it's helpful to check out invasive species websites to get an idea of what you're looking for. But even more important is to look for areas in your woods that seem to be dominated by a single understory plant. Pick it, identify it, and if it's undesirable, then poison or pull it.

Getting rid of invasive species is an ongoing project. We dedicate certain periods throughout the year to dealing with prickly ash (a native species that becomes invasive on compacted soils), which took over large patches of our woods after they'd been pastured for many years. Prickly ash is similar to buckthorn, which is a major problem in woodlands in much of the country. Over the years we've fine-tuned our eradication methods to be both organic and effective, if rather labor intensive. From late fall through winter we cut off the tops, so the trees are about thigh level, and burn the tops (one of the few good things about prickly ash is that it's oily and burns easily when green). In early spring when the ground is soft, we pull the roots. We pull small trees by hand; we use a mechanical device for the bigger stuff and an electric winch for the monsters. We then stack and burn them the following winter. For the next two or three years, we patrol the area and pull the resprouts. You could instead cut off the tops of woody invasives in late summer to early fall, when the plants are moving stores to their roots for winter, and dab poison on the stumps. This will kill most of the roots.

Wildlife

Most woodland owners enjoy and want to encourage wildlife on their acres. Provide food, shelter, water, and space according to the needs and preferences of desired species, and they will probably come. Young woods and brushy edges — where woods meets grassland or water — are excellent sources of shelter and food for many animals, and they are especially attractive to many game species such as deer, rabbit, and grouse. Mature woods with little ground cover tend to attract many species of songbirds and tree-dwelling mammals. Turkeys favor healthy oak woods that produce a lot of acorns. Manage your woods for a diversity of tree types and ages, and you'll have a diversity of wildlife.

A managed woods can be very beneficial for wildlife — if you're a deer hunter, this can improve your success rate.

Equipment and Safety

The basic tool most landowners need for woods projects is the chain saw, and the best thing you can do for yourself after buying one is to take a chain saw safety course (usually a one-day commitment) to learn how to use it efficiently and safely. Chain saw accidents are discouragingly common and generally preventable, not to mention expensive to patch up and potentially disfiguring, disabling, or fatal. Courses may be sponsored by Extension offices, woodland owners organizations, or logging associations.

Purchase the smallest saw suited for the work you intend to do, and your shoulders and back will thank you over the years. Also invest in ear muffs, a hard hat, safety chaps, and steel-toed boots for protection. Loppers for pruning, a cant hook or similar tool for rolling and positioning logs, and wedges to hold cuts open and lean trees in the right direction are enough to start with. A brush mower of some type is nice for mowing trails.

Harvesting your own timber presents the problem of getting it sawed into boards and beams. In many areas where there are small woodlands and private owners, there are people who will process your logs. You may even find an entrepreneur with a portable mill who will come to your land so you don't have to haul the logs anywhere.

Timing Your Woods Work

Do woods projects when the soil is dry or frozen. This prevents soil compaction, erosion, and rutted trails. It's dangerous to work in rainy periods, as leaves, mud, and logs are all slippery when wet.

Freezing temperatures also prevent disease transmission, which is a major threat in woodlands throughout the country. When possible, do woods work in winter. If winters are wet instead of frozen in your region, save woods projects for dry spells and be cautious with any equipment that could throw a spark that might start a fire.

Planting Trees for Woodlands

For large numbers of seedling trees on relatively flat ground, you may be able to rent or borrow a planting machine that you can pull behind your tractor. If you're on rough ground, or if there are no machines available in your area, you can plant small trees quite quickly by hand.

While you're planting, it's essential to keep the roots moist (not wet — put them in damp sawdust or hay, not a bucket of water) and out of the sun. This is as important for success as getting them in the ground properly.

The planting technique is simple: Jam a sharp shovel deeply into the soil and lever the handle end of the blade forward to create a wedge-shaped opening in the ground. Put the tree in the hole, making sure the roots go straight down and aren't bent or curled at the bottom. Jam the shovel into the soil a few inches from the original opening, move the bottom to firm the soil around the roots of the sapling, then move the top of the blade to close the slit around the sapling. Remove the shovel, give a good hard stomp to close the second slit and further firm the soil, and you're ready to plant the next tree.

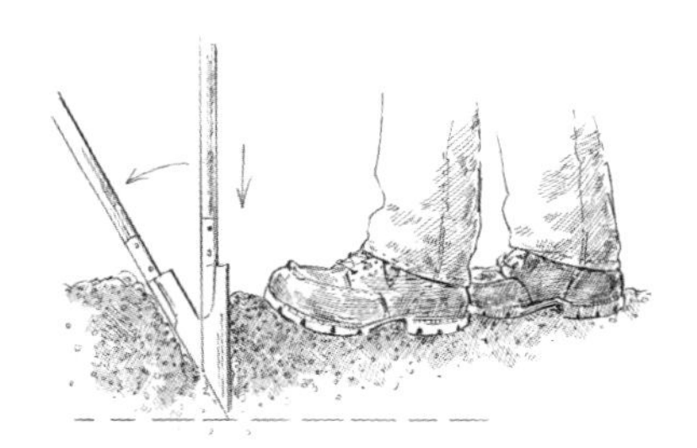

1. Jam and lever.

2. Place roots in hole.

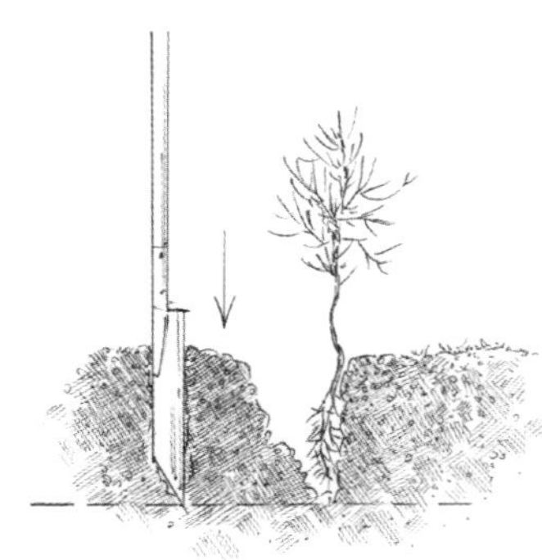

3. Make second jam.

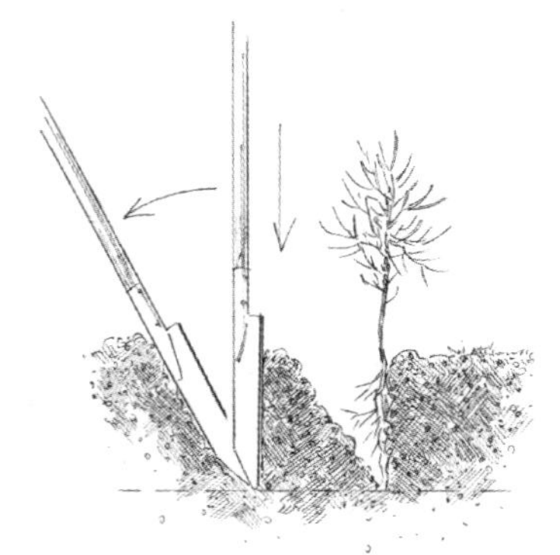

4. Lever dirt back to close first hole over roots.

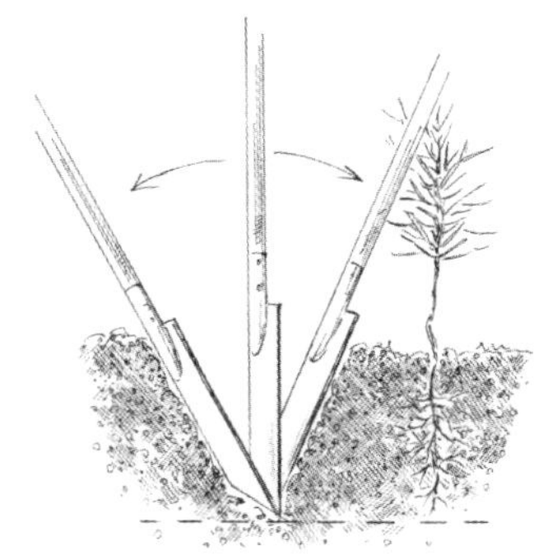

5. Make sure dirt is well packed.

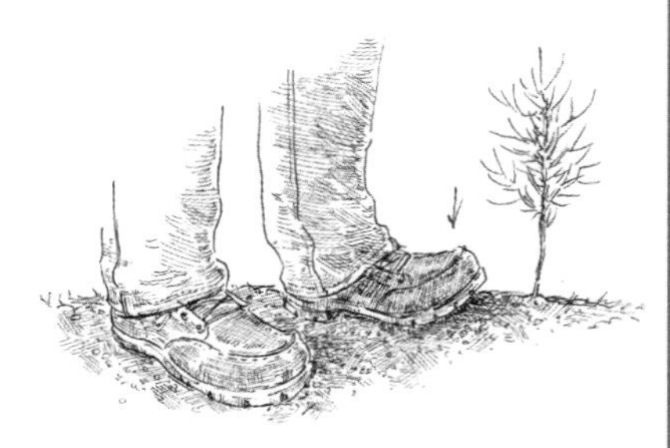

6. A final stomp finishes the job.

CHAPTER ELEVEN

Late Fall

Daytime soil temperature falls below 50°F/10°C; nighttime freezing air temperatures are frequent.

GRASS GROWTH STOPS and nights are regularly below freezing in late fall, the traditional season for killing hogs, cutting firewood for the following year, and deer hunting.

The weather can turn wet and cold for days at a time, and you never know how quickly winter will arrive. The corn harvest continues as fast as possible, until it is completed or stopped by bad weather. It's time for triage: finish what must be done before cold weather, and the rest will have to wait until spring if you run out of time. Store a pile of wood scraps and dry twigs in the barn for woodstove kindling on wet days, and move seasoned firewood into the covered woodshed, where it's handy to the house and will stay dry.

Late Fall

FOCUS ON:
pigs • woodlot

Garden

» **Finish putting the garden** to bed.

» **Clean, sharpen, and store tools.**

» **Give the compost piles** a final turn before they freeze up.

Field

» **Leave crop residue on the surface** to protect the soil over the winter. It's too late in the season to plant winter cover crops.

Pasture

» **Continue to take down temporary electric fencing** as livestock graze each paddock for the final time.

Orchard

» **Finish up whatever hasn't been done yet** to prepare for winter.

Beeyard

» **Wrap colonies with tar paper** or an insulating wrap in cold areas.

Barn

» **Move livestock into winter quarters** and feeding areas as grazing ends.

» **Butcher the pigs** or send them to the processing plant ahead of deer hunting season.

» **Breed does and ewes now** for April kids and lambs.

Coop

» **Hang a heat lamp** if desired.

» **Check ventilation;** there should be fresh air moving through, but no bad drafts.

Equipment Shed

» **Place any equipment that will be used during winter** near the shed door for easy access.

Woodlot

» **Start or continue cutting firewood** and thinning and pruning trees.

Wildlife Habitat

» **Build brush piles** and tangles with debris from pruning, thinning, and making firewood.

» **Leave some big logs on the ground** to act as drumming logs for grouse and shelter for amphibians.

Garden

Except for those few very hardy vegetables that are still producing, like kale and parsley, this is the time to finish putting the garden to bed for winter, as discussed in the previous chapter. Cover crops should be up and dead or dormant, and mulch should be spread where no crops were planted. Except for gradually covering the garden with a sprinkling of wood ashes as they are available, there's nothing to do until it's time to start those earliest seedlings indoors or in cold frames in late winter. Give the compost piles a final turn before they freeze up, and take a rest.

Field

If you haven't planted your cover crops, then leave the crop residue left after your harvest to protect the soil through winter. As with the garden, there really isn't anything else to worry about at this season in the fields.

A fall-planted cover crop (here in a cornfield) protects the soil from winter winds and water erosion.

Pasture

Grazing livestock may still be finishing their final rotation through stockpiled pasture paddocks and hayfields. Take down and store the temporary electric fence as they finish each paddock (but only when it's dry — wet wire will rust in storage). If you have a lot of legumes in your pastures and hayfields, don't graze or mow too short. An insulating layer of several inches of stubble protects against frost heaving in late fall and early spring, which can injure and kill these plants. Once the last paddocks have been grazed, move the livestock into winter quarters and let the pastures rest through winter.

This is also a good time of year to mow or cut off overgrown vegetation that is on permanent fences (I always find myself pulling wild grapevines off the wire), which greatly increases the lifespan of your permanent fences. Wildlife habitat is fine alongside fences, but not on them.

Orchard

Finish preparing the orchard for winter, as discussed on page 150.

Beeyard

Prepare hives for winter. Insert an entrance reducer if desired to protect against unwanted intruders, especially mice. In cold areas, wrap hives in black tar paper for insulation. You can add insulation under the cover, but leave an opening for ventilation to prevent moisture and CO_2 buildup. In windy locations, put a rock or a strap on the cover.

Barn

As the weather continues to cool, it's an excellent time for larger-scale home butchering (sheep, goats, and pigs). If you are planning on shipping the animals to a processing plant instead and haven't yet done so, make an appointment before deer hunting season or you may have to wait quite a few weeks to get your meat back.

For April kids and lambs, breed does and ewes now.

For a smooth transition to winter feed, keep hay available for livestock at all times as the grazing season winds down. Leave it in the barn or barnyard, or open up the winter feeding area while still keeping the gates open to the last grazing paddocks. If you are wintering livestock in the barn and barnyard, the NOP Final Rule mandates and common sense dictates that they have access to the outdoors when weather permits, which in reality is most of the time. Open up an adjacent paddock for their winter use. Sunshine and fresh air are great boosters of herd and flock health, and time outside means less manure and soiled bedding to haul out of the barn and coop over the winter.

Move all remaining livestock into their winter quarters, or, if summer and winter quarters are the same, make sure pens and sheds are well ventilated but draft-free. Pigs, poultry, and goats especially need draft-free, well-bedded housing. Cattle, horses, and sheep generally do fine in open-sided sheds, with the open side placed on the south, for the sunshine.

All animals need access to unfrozen, clean water every day year-round. From now until spring, you can break ice on water tanks twice a day, replace frozen water with thawed every day, or put plug-in heaters in the water (available at any farm store or through farm catalogs), which, like heat lamps, add to the electric bill but are very convenient. In freezing weather, you can't automatically fill stock tanks by using a hose hooked to a float valve, so tanks must now be filled by bucket, hose, or, if you're really fancy, an underground insulated line to the water tank.

Even grazing livestock — most often cattle and sheep — that are outwintered (fed hay bales on open pasture or field through winter) need to have access to unfrozen water; don't believe the old-timers who tell you they'll do fine eating snow. They also need a place where they can be dry and out of the wind in bad weather. Build a lane (temporary electric fencing works fine) or leave gates open so livestock can always get back to the shed and the water tank. In areas where there's snow but very little cold rain, cattle don't need a roof as long as they have a windbreak.

Coop

The laying hens should be tucked in for winter. Leave the coop open daily except in bad or extremely cold weather, so they can get a little sun.

Chickens need a draft-free coop in winter.

Mud Season Reprise

When the soil becomes soggy instead of frozen as you head into winter, the resulting mud (as in early spring) makes livestock care a little trickier. A long mud season can really put a monkey wrench in the works.

Soft, wet soil is rapidly churned into mud by hooves, which compacts and destroys soil structure. Plus it's hard on the animals: standing up to their ankles in cold mud is chilling, and it's harder to move around when they have to yank their hoof free with every step. If there's nowhere else to lie but in the mud, your animals will be muddy all over, which mats down their coats and cuts the insulating value, making them colder still. Plus mud is just difficult and dirty if you're trying to move feed, animals, or equipment through it.

If livestock have access to a dry shed, they can get out of the mud when they need to lie down. A cement floor in the barnyard, especially around the water tank, is a good investment for minimizing mud season.

Equipment Shed

As you finish with the last pieces of equipment for the season, clean, fix, tune up, and store them. Then pull out winter equipment — for us that means the chain saws, log splitter, and snow blower — and make sure they start and run. If you don't want to store the power washer in the basement where it won't freeze, then put antifreeze in the waterline (see the owner's manual for instructions).

Woodlot

For those who have a woodlot, late fall is the best time of the year to be in it. Other work is tapering off, the bugs are gone, the air is cool, and the leaves are down so you can see the form of the trees and the lay of the land. If you don't own any woods but could use some firewood, ask neighbors and friends who do have woods — many woodland owners are happy to give you the firewood in return for your help in cleaning up downed and crippled trees.

Though the year's firewood supply is traditionally cut in mid to late fall, you can harvest firewood any time of year, and certainly all winter as you do your woods cleanup work. Make firewood from freshly downed trees, or cut the standing stunted, crowded, crooked, cracked, dead, dying, or diseased trees for firewood and leave your straight, healthy big trees for beauty, wildlife, and timber. But leave a few big dead, standing snags per acre for critical wildlife habitat — many dozens of species of insects and wildlife depend on these snags for food and/or shelter. For days when weather keeps you out of the woods, keep a supply of rounds in the shed or near the woodpile for splitting into firewood.

Wildlife Habitat

As you cut firewood and thin and prune trees, use some of the debris to build brush piles for wildlife, as discussed on page 157. Game cameras are an excellent tool for observing how animal patterns change as winter arrives, such as where they're finding winter food and shelter; you can use this information to plan for enhancing winter habitat.

Big trunks that have been on the ground for a while and are getting too punky to make good firewood should be left for drumming logs for grouse, especially if they're hollow and near brushy cover. These logs are excellent shelter for small amphibians, too, including salamanders.

Making Firewood

Firewood is best made from standing crippled or unhealthy trees and freshly downed trees; this cleans up the woods and spares the good trees for timber and wildlife. Fell the tree, cut off the branches too small to be of use (a judgment call, for sure), then cut the remainder into woodstove lengths. You can split it on the spot, or haul it back to the woodshed or yard for splitting later. We keep the splitter in a little area behind our covered woodshed, pile the rounds nearby, and finish the splitting as time allows. When our three sons were home and our shoulders a lot younger, we split everything by hand and enjoyed doing it. These days we use a hydraulic splitter — noisy, but very efficient.

Before you fire up the chain saw, make sure you have proper safety equipment and have taken a safety course or learned from someone who knows how to use a saw correctly (see page 159 for more information). Once wood is cut and split, stack it and let it dry for at least a year before using it for the stove. In our cool, moist climate, we cure our wood for two full years before burning it, so that we always have three wood piles going — one just moved into the woodshed for this winter, one in the open for next winter, and one we're building now for the year after next. An outdoor furnace will burn green (freshly cut) wood, and in dryer, warmer regions, wood for indoor stoves may only need to be cured for a year.

Deer Hunting Season

Though the timing of deer hunting season varies considerably around the country, in general it occurs during late fall or early winter. Besides potentially putting some venison in the freezer, we've found it to be a wonderfully social time in a rural neighborhood, as the hunters visit around to make their plans and share their stories.

If you're interested in supplementing your larder with some locally raised venison (or any other game animals) but haven't ever hunted, you will be required to have both a license and a hunter's safety certificate, which is gained by taking a brief, interesting course offered by your state's Cooperative Extension Service or Department of Natural Resources. Then acquaint yourself with an experienced and ethical hunter to learn the art of the chase.

If you don't hunt, I highly recommend giving permission to friends, relatives, and/or neighbors that you know and trust to hunt your land if you own enough land to support hunting. This tends to keep unknown hunters out, builds good neighborly relationships, and is good for the woods: the overpopulation of deer in most areas is degrading forests.

Regardless of whether or not you hunt, if you are in the woods at this season you should wear a blaze orange hat and shell or jacket. And put away the white handkerchiefs and Kleenex — they look too much like the flash of a deer's tail. Have something red or blue in your pocket instead.

Topic of the Season

Restoring Biodiversity

Biodiversity (a variety of plant and animal life) is essential to the resilience and so the long-term survival of the ecosystem. A healthy ecosystem fosters living fertile soil, air and water purification services, and the recycling of dead things into new life.

Resilience is the ability to recover quickly from inevitable natural disturbances like fire and disease and is the foundation of sustainability. One way to illustrate the importance of resilience is to compare how the Corn Belt states looked prior to European settlement with how they look today.

Driving through heavily farmed areas of the Corn Belt is above all monotonous. In summer the fields are full of corn and soybeans and nothing else, and in winter they're bare. The synthetic fertilizers and pesticides not only kill the target plants and insects but others as well, and they degrade soil life and structure.

A woodland pond expands the number of species that can make a home in your woods.

Less life in the soil means poorer and slower breakdown of dead plants and animals, so that nutrients and organic matter aren't so quickly absorbed and so not completely recycled. With less organic matter, the soil can't absorb and retain as much water, either. Water runs off the surface year-round, carrying topsoil, pesticides, and fertilizers with it, or leaches through to the groundwater table, carrying nutrients along — as well as synthetic fertilizers and pesticides. These are significant pollutants in many rural surface waters, wells, and municipal water systems. Continually growing the same crop provides a nice stable habitat for the pests and diseases that feed on that crop, while the insects and animals that would feed on and provide some control of the pests don't have a place to live and nothing to eat for half the year when the crops are gone. One pesticide-resistant strain of insects or disease or weed, an increasingly common phenomenon, can decimate the crop.

Contrast this with the abundance of plants and animals and the fertile soils of the original prairies and woods. Year-round diverse plant cover protected the soil from erosion. Under the surface, soil-dwelling organisms had a continual and varied diet of plant and animal matter to digest and recycle, which built organic matter and fertility that supported a highly diverse aboveground community of plants and animals. Nothing was wasted; whatever was produced was consumed and eventually progressed back into the soil to nourish new life. Nutrients were conserved, and if some members of the community were destroyed by disease, fire, or other natural disasters, there was always another to occupy that niche in the ecosystem. The many pests, pollinators, predators, and prey kept each other in balance over time. Diversity preserved resiliency.

Raising plants and animals for human use by its very nature decreases diversity. We are occupying ecosystem niches and diverting resources for our own purposes, displacing the original inhabitants and disrupting nutrient cycling. If we wish farming and the human life that depends on it to continue into the future, it behooves us to do whatever we can to maintain the healthy soil that is the foundation of productive farming in the long term. It also behooves us to maintain, as much as we can, the diversity of plants and animals that are native to our land. By doing this, all the parts — insects, prey and predator animals, even diseases — stay in balance, so none gets out of control.

Promoting ecological balance and conserving biodiversity is done bit by bit through the year and over the years, just like most other chores on a rural acreage. The steps involved are:

1. Research what your land looked like before settlement: what plants and animals were present? This is often a winter job, when there's more time for reading.
2. Identify and preserve what's left of the original inhabitants. This is a year-round job, especially since you will need to recognize plants in different stages of growth. Identification should include reading up on what the preferred habitat and foods are for that species.

3. Set aside areas to build or expand habitat for native species. This could be as simple as leaving strips of land for native grasses, forbs, and shrubs along garden, field, and orchard edges. Or it could be as ambitious as restoring a wetland or prairie, or as specific as extirpating an invasive species in your woodlot and reestablishing the natives that have dwindled or disappeared. Wetland work is best done during dry spells, and while native species are often planted in spring, removing the current vegetation generally requires smother crops and repeated tillage during the previous growing season.
4. Build habitat in many small ways. Construct brush piles and rock piles in the woods and along brushy edges; put up birdhouses and bat and bee houses; leave a trickle from the hose going when the swallows are building their nests and need the mud; leave some grasses by the edge of your little pond long, so frogs, turtles, dragonflies, and other small life can seek shelter there.

People choose to live in the country for many reasons: to have the pleasure of growing their own food, for privacy, for space, and to enjoy a beautiful landscape. Of the many good things that happen when we preserve and build biodiversity on our land, simply being able to live as part of a natural community bursting with life might be the happiest and most satisfying of all.

CHAPTER TWELVE

Early Winter

Soil temperature falls below 40°F/4°C. Plant growth ceases.

JUST WHEN YOU THINK the hibernators have finally hibernated, you'll have a warmish evening and you'll smell a skunk or see a late moth. Early winter can flicker in and out for a while; snow falls then melts; ponds freeze hard then slushy spots appear. But once the ground starts freezing, there's no doubt that it's winter. It's time to focus homestead projects on the woods and on inside pursuits.

Take advantage of snow. If there's an ample amount and it lasts long enough, you can pile snow around foundations to stop drafts, over tender perennials to insulate the roots from the tearing effect of freeze-thaw cycles, and over "clamps" of vegetables (root crops stored in trenches under straw and soil) in the garden as added insurance against freezing. Snow also gives you a great opportunity to see what the wildlife is doing and even where your livestock is hanging out when you're not around. While game cameras take snapshots, tracks tell entire stories. But it's a rare place now that has reliable snow cover throughout winter. If the ground is bare, place some straw, old hay, or leaves on top of water-lines and perennials to insulate them.

Early Winter

FOCUS ON:
woodlot

Garden

» **Start early, hardy plants** under cover in mild climates.

» **Sprinkle a light dusting of wood ashes** on the soil.

Field

» **Rest time.**

Pasture

» **Remove livestock from all pastures** except for outwintering areas.

» **Continue checking and repairing fences** until cold weather or deep snow makes it too onerous.

Orchard

» **Survey occasionally** to make sure rodent and deer protection is staying in place.

» **Sprinkle wood ashes** around trees at the dripline.

Beeyard

» **Hope the bees will winter well.**

Barn

» **Settle into a winter routine** of daily feeding, watering, and checking.

» **Poison rats and pigeons** as needed.

Coop

» **Collect eggs mid-morning** to get them before they freeze.

Equipment Shed

» **Keep chain saws clean** and blades sharpened.

Woodlot

» **Thin and prune trees** and harvest timber, except where winters are wet rather than dry or frozen.

Wildlife Habitat

» **Put out mineral and salt licks** for wildlife.

Garden

As you accumulate wood ashes from the stove, spread them with a light hand over the garden and flower beds. This adds potassium (potash) to the soil.

Field and Pasture

Rest time.

Orchard

When you've spread enough ashes in the garden, start sprinkling them around the dripline of trees.

Beeyard

There is very little to do at this season, except to hope the bees are wintering well and savor the honey you gathered earlier in the year.

Barn

By now the herd or flock is settled in winter quarters, and all you need to do is feed, water, and check them daily for problems — especially shivering and other abnormal behavior that can indicate discomfort or illness. Make sure that all are staying dry, eating, and drinking water (sometimes a heater can develop a short, putting a charge on the water, and animals will quit drinking). Watch for icy spots in lanes and around water tanks; apply sand, ashes, or scraps of straw or old hay to make them less slick. Continue freshening salt and mineral feeders. Keep an eye on the weather forecast — it's easier to feed and bed a little earlier than usual rather than in the middle of a howling storm!

Many beekeepers in northern areas wrap their hives in the winter so it's easier for the bees to keep warm.

Winter Manure Handling

There are a few ways you can handle manure in winter. If you're wintering livestock outdoors on pasture or field, there's no handling required — the animals are spreading the manure themselves as they move through the bales you set out last fall. Poultry can be deep bedded; just put some more bedding on top of the old stuff when it gets dirty, then clean it all out in spring. This works well for a few chickens, but if you have a lot of chickens, it's a lot of work! In that case it's best to do regular clean-outs throughout winter. You can deep-bed livestock, too, but you may need a skid steer to do the clean-out in spring. The more common approach is to clean indoor stalls and pens regularly and neatly pile the manure where you can get at it easily in spring for spreading. If you're in a region where the ground is bare and dry through at least part of winter, you can spread manure in winter if there's no danger of its being washed into a nearby body of water.

If you won't be spreading manure and dirty bedding immediately, form it into tight piles so it will compost. You can turn the piles with a shovel or skid steer, but that isn't necessary.

Coop

Hens will continue to lay through winter, though not as much. A heat lamp will bump production. Collect eggs by mid-morning, so they don't freeze.

Equipment Shed

While all of the maintenance and repairs you've done are still fresh in your mind, make notes on what you've done, what broke, and what parts you need for next year.

Woodlot

With the rest of the place on autopilot for winter, there's time to work in the woods, pruning, thinning, making firewood, harvesting timber, and burning debris. In winters when the snow gets deep, you can usually keep trails open with snowshoes or by repeatedly driving the ATV over them. I've even used a pulk (a sturdy sled) to haul equipment.

Wildlife Habitat

Mineral and salt licks are appreciated not only by deer but by other mammals as well. It is at best not sporting and at worst illegal (check your state regulations) to shoot deer over bait. Licks may not be technically classified as bait by your state regulations, but the idea is the same. Wait until hunting season is over to set these out.

Brush and other woody debris can be burned to reduce fire hazard, improve aesthetics, and minimize some disease and pest problems. Burn only when the ground is snow-covered or wet.

Topic of the Season

Quality of Life

The harvest is over. It's the end of the year. This is traditionally a time for stepping back and reflecting, and also for celebrating family and friends. Raising good food is interesting, satisfying work; eating it is delightful and healthy. And yet, there is so much more to life.

We are part not only of a natural ecosystem but of a human one as well, consisting of friends, families, neighbors, our local communities, and finally the wider world. Our acres do not exist in a bubble; they are an integral part of a wider ecosystem that both influences and is influenced by how other people are using or abusing their land and the water and air we all share. We don't live in a bubble, either: to a lesser or greater extent, we depend on roads, utilities, police, fire and ambulance services, schools, churches, retail stores, farmers, and manufacturers.

Unless we're dedicated hermits, we also rely on other people for fellowship, fun, and entertainment. Scheduling time to participate in local affairs adds balance and fun to life, and it builds resilience in the community. The more people care about what happens locally and get involved in making things happen, the more likely that a community will thrive.

There are two general categories of involvement: (1) groups and organizations that are involved with taking care of the community, and (2) stuff that's done for fun and sociality. The former includes local government bodies responsible for such things as land use regulation, road maintenance, and schools, as well as organizations like churches and other charitable groups that are proactive in addressing local social needs such as poverty, care for the elderly, and substance abuse. The latter category includes summer softball and trap shooting leagues, weavers' guilds, farmer and woodland owners' organizations, community theater, choirs, and bands.

All of these activities involve meeting regularly with others in the group to discuss, rehearse, or play. Government bodies — town boards, planning commissions, and various committees — tend to meet year-round, as do church committees and organizations like the Rotary Club.

Tracking Your Year

For planning and scheduling, it's very useful to think through the annual cycle of each part of the homestead. For example, the orchard moves from requiring pruning to blossoming to fruit drop, and then to harvest time and needing to be put to bed. In this section, we look at the annual cycles of garden, field, pasture, orchard, beeyard, barn, coop, equipment shed, woodlot, and wildlife habitat as they all dance together through the seasons.

Throughout the Year

Garden

» **Midwinter.** Read your notes from the past growing season to assess how well the previous year's varieties, rotation, and cover crops performed. Use this information to plan this year's crop rotation and cover crop plantings, then order seed. If soil amendments (based on soil test results) were not applied last fall, price and order/purchase them for spring application.

» **Late Winter.** Get out containers, starter soil mix, lights, and other seed-starting supplies. Seeds requiring the longest growing seasons can be started indoors. If you're craving salad, pull the mulch off a corner of the garden, add a cold frame, and plant early greens.

» **Early Spring.** Peas and other cold-tolerant vegetables can be planted as soon as the soil softens. Spread compost and purchased amendments (such as lime and blood meal) on the rest of the garden when the soil begins to firm up. Till as soon as the soil is dry and firm enough to work. Start more seed inside for mid- and late-spring plantings.

» **Mid-Spring.** It's primary planting season. As the soil temperature rises, seeds go in the ground according to their temperature preference. Rhubarb season.

» **Late Spring.** As the soil warms to 70°F/21°C degrees and higher, plant the most tender seeds and begin succession plantings. Control weeds early and often. Pick early vegetables. Transplant most indoor seedlings to the garden, but have covers ready for the tomatoes, peppers, and squash in case of frost.

» **Early Summer.** Continue succession planting and stay on top of the weeds. Mulch if it's dry, water as needed, and enjoy early vegetables and herbs. It's strawberry season.

» **Midsummer.** Weed pressure lessens and harvesting accelerates. Keep watering. As vegetables are harvested, sow fall crops, such as garlic, carrots, parsnips, rutabaga, cabbage, kohlrabi, and salad greens — anything that will mature in half a growing season and tolerate cool fall weather. Test your soil.

» **Late Summer.** Harvesting continues. Sow the last succession crops such as spinach and lettuce.

» **Early Fall.** The harvest winds down. As rows and beds are vacated, apply compost, manure, and other amendments before tilling and replanting with cover crops or covering with mulch. In warm areas, plant hardy vegetables that will overwinter a couple of weeks before first frost, and protect them with row covers or straw. Plant garlic.

» **Mid-Fall.** Put the garden to bed. Finish cleanup, spread compost and manure, till if you like, and cover soil with mulch if you didn't plant cover crops.

» **Late Fall.** Finish putting the garden to bed. Clean, sharpen, and store tools. Give the compost piles a final turn before they freeze up.

» **Early Winter.** In very mild climates, you can start early, hardy plants under cover. Sprinkle a light dusting of wood ashes on the soil.

Field

» **Midwinter.** Read your notes from the past growing season to assess how well the previous year's varieties, rotation, and cover crops performed. Use this information to plan this year's crop rotation and cover crop plantings, then order seed. If soil amendments (based on soil test results) were not applied last fall, price and order/purchase them for spring application.

» **Late Winter.** Receive and store seed away from moisture and rodents. Manure can be spread while the ground is still frozen if there is no danger of runoff during snow melt or spring rains.

» **Early Spring.** As green-up begins, check for winterkill in the hayfields. When the soil can be worked, reseed any affected areas at the same time that you disc and overseed winter feeding areas. In the crop fields, as soon as the soil is dry enough to run equipment without rutting or getting stuck in the mud, spread manure (from winter feeding areas and the barn), then till under both manure and winter cover crops. Pick rock. Do finish tillage, such as discing or dragging, to prepare a good seedbed. Plant spring small grains as soon as possible.

» **Mid-Spring.** Plant row crops.

» **Late Spring.** Finish planting. Start cultivating row crops when they're a few inches high.

» **Early Summer.** Make the first cutting of hay. Continue to cultivate row crops as needed until the leaves of the plants in adjoining rows meet in the middle and shade the ground, suppressing weeds.

» **Midsummer.** Small grain harvest begins. Row crops should be high enough by now to shade weeds, so cultivation ends. Test your soil. Spread manure and lime on cut hayfields.

» **Late Summer.** Cut hay. If weeds were a problem in the small grains, this is an opportunity for trying a "stale bed" method of weed control before planting forages or winter cover crops.

» **Early Fall.** Winter small grain planting begins. Vacant fields should be planted with winter cover crops.

» **Mid-Fall.** The first hard frosts kill the corn, and harvest begins (for grain corn; silage corn is already done).

» **Late Fall.** It's too late in the season to plant winter cover crops; leave crop residue on the surface to protect the soil over the winter.

» **Early Winter.** Rest time.

Pasture

» **Midwinter.** Order seed for overseeding winter feeding areas in early spring. If there's little or no snow, work on clearing fence lines of brush and branches.

» **Late Winter.** Move livestock off wintering areas to the barnyard or to sacrifice paddock before soil gets soft. Frost seed poor areas and winter feeding area. Clear brush and branches off permanent fences and repair any broken wires and posts.

» **Early Spring.** Finish clearing and repairing permanent fences. Start putting up temporary fencing for rotational grazing. Disc and seed poor areas if they haven't been frost seeded. Grazing can start as soon as the ground is dry enough that hooves won't cause serious damage.

» **Mid-Spring.** Forage growth accelerates. Move livestock through large paddocks as this occurs, letting them just take off the tops to keep grasses and clovers from maturing too quickly.

» **Late Spring.** There is explosive grass growth. Finish putting up temporary fence and keep the animals moving. Control weeds by mowing. If the grass starts growing faster than the livestock can keep it grazed, set aside some paddocks for making hay and speed up the rotation.

» **Early Summer.** Monitor paddocks for slowing grass growth and adjust grazing rotation accordingly. Continue weed control. Check temporary fences often to make sure they're up and powered. Mow fence lines.

» **Midsummer.** As the weather gets hotter and dryer, growth of cool-season forages will continue to slow. Slow down rotations to give paddocks adequate recovery time, or add paddocks, or shift the rotation to paddocks with warm-season forages. Clip paddocks after grazing, if needed, to control weeds. Test your soil.

» **Late Summer.** Adjust rotation as appropriate for rate of grass growth; continue weed control.

» **Early Fall.** As temperatures begin to cool and moisture levels increase, cool-season grasses grow rapidly again for a few weeks. Warm-season grasses shut down for the year. Adjust your grazing rotation as indicated. Continue weed control and mow fence lines again if you find the time. Stockpile paddocks for late fall and early-winter grazing.

» **Mid-Fall.** Take down temporary electric fencing for the season as livestock graze each paddock for the final time.

» **Late Fall.** Continue to take down temporary electric fencing as livestock graze each paddock for the final time.

» **Early Winter.** Livestock should be off of all pastures except for outwintering areas. Continue checking and repairing fences until cold weather or deep snow makes it too onerous.

Orchard

» **Midwinter.** Select and order new trees and pest controls — traps and treatments — if you want them. Sprinkle wood ashes around trees at the dripline.

» **Late Winter.** Prune. Remove any tent caterpillar egg cases you see and assess winter damage from rodents, deer, sun, and disease.

» **Early Spring.** Finish pruning before green-tip bud stage, remove the debris, and decide whether to save or replace damaged trees. Plant new trees. Continue to look for and destroy tent caterpillar egg cases and nests.

» **Mid-Spring.** If you intend to actively control pests and diseases, start monitoring now for problems by observing trees, measuring degree days, and placing pest-specific traps.

» **Late Spring.** It's apple blossom time! Chemically thin blossoms if desired and continue to monitor for insects and disease.

» **Early Summer.** Mow. Thin fruits when they're 1 to 1½ inches in diameter, if desired (I never have enough time). Clean up drops. Pick red currants, gooseberries, and strawberries.

» **Midsummer.** Continue to monitor for pests and disease. Test your soil. Peaches, plums, apricots, both sweet and sour cherries, raspberries, and huckleberries are in season.

» **Late Summer.** Mow. Pick blueberries, black currants, pears, early apples, and early grapes.

» **Early Fall.** Mow the orchard again. The primary apple and grape harvests begin.

» **Mid-Fall.** Finish the apple and grape harvests, and put the orchard and vineyard to bed for winter.

» **Late Fall.** Finish up whatever hasn't been done yet to prepare for winter.

» **Early Winter.** Walk through occasionally to make sure rodent and deer protection is staying in place. Sprinkle wood ashes around trees at the dripline.

Beeyard

» **Midwinter.** Observe hives on warm, sunny days (over 45°F/7°C), looking for bees emerging for "cleansing flights." Do not open hives. Order hives, supplies, and bees for spring delivery. Assemble and paint hives when they arrive.

» **Late Winter.** As in midwinter, check for activity level in hives (but don't open) on warm days. Feed.

» **Early Spring.** Continue to feed if indicated up until the first major nectar flow. Open hives on a warm day to check for health and strength. Medicate according to regional recommendations, clean up any winter mouse damage, and replace worn combs and parts. Place new hives, and hive new queens and bees.

» **Mid-Spring.** The biggest nectar flow of the year is getting underway as fruit trees and many other plants blossom. New hives may still need feeding until well established. Observe hives regularly, but leave them alone unless a problem is evident.

» **Late Spring.** Add supers as needed.

» **Early Summer.** Add supers as others fill up; filled and capped frames can be taken for honey.

» **Midsummer.** Primary honey harvest begins. Leave a full super for bee food and supply hives with more empty supers.

» **Late Summer.** Finish the first honey harvest. Add supers as needed.

» **Early Fall.** The last honey harvest happens now. Leave enough to feed the bees in winter.

» **Mid-Fall.** Feed syrup, if desired, to ensure a good honey supply for the hive through winter. Install entrance reducers to help keep cold out.

» **Late Fall.** In cold areas, wrap colonies with tar paper or an insulating wrap.

» **Early Winter.** Hope the bees will winter well.

Barn

» **Midwinter.** Check the hay supply: you should have at least half left. If you're short, start looking to buy more. Check all livestock and water, salt, mineral, and hay feeders daily. If desired, ultrasound goats and sheep to see how many young they are carrying. Control rodents. Research where you can buy weaner pigs in spring.

» **Late Winter.** Prepare pens for any livestock that are due to calve, lamb, or kid later this season or in early spring. Four to six weeks before lambs are due, ewes and does should be vaccinated, introduced to or given increased grain rations, and "crutched" (shearing or trimming butts). If you plan to raise weaner pigs this year, lock in a spring purchase from a pig producer.

» **Early Spring.** Finish winter manure cleanup. Finish spreading or stack or compost the manure. Keep livestock penned in the barnyard or sacrifice area until the soil dries and firms up. Calving, kidding, or lambing begins for most. Prepare sty or hut and paddock for weaner pigs, and lock in an agreement for picking them up after weaning.

- **Mid-Spring.** Calving, lambing, and kidding start or continue. Buy and bring home weaner pigs. Castrate and vaccinate lambs and kids. Calves can be castrated now or in the fall. Arrange to have sheep sheared by early summer.

- **Late Spring.** All of the livestock are out on pasture. Calving, kidding, and lambing finish. Find a bull or an artificial insemination service for breeding the cows.

- **Early Summer.** Young livestock should be getting the hang of the pasture rotation by now. Offer shelter during the day if animals are badly bothered by flies.

- **Midsummer.** Cull infertile or problem animals. Start meat animals on grain if desired. Turn out the bull. Make sure shade or shelter is available on very hot days. If flies are really bad, let animals in during the day and out to graze at night.

- **Late Summer.** Ready facilities for weaning and other upcoming stock work.

- **Early Fall.** Buy hay if you're short. Make sure you'll be able to move hay to animals or animals to hay in winter bad weather. Wean young stock; castrate and vaccinate calves if you haven't already done so.

- **Mid-Fall.** Finish preparing winter quarters and feeding areas, making sure you'll be able to move hay to animals or animals to hay in bad weather. Put heaters in water tanks. Ship any animals you don't plan to butcher or carry through winter. Pigs may be butchered or shipped now, or next season when it's cooler and they're heavier.

- **Late Fall.** As grazing ends, move livestock into winter quarters and feeding areas. Butcher the pigs or send them to the processing plant ahead of deer hunting season. For April kids and lambs, breed does and ewes now.

- **Early Winter.** Settle into a winter routine of daily feeding, watering, and checking. This is also a good time of year to poison rats and shoot pigeons.

Coop

» **Midwinter.** Order chicks for delivery in early to mid-spring. Check, feed, and water hens daily. Let them outside on nice days.

» **Late Winter.** Get the brooder ready for chicks. Clean out manure and bedding from the hen coop and spread or compost it.

» **Early Spring.** Set up the brooder and get the temperature right several days before new chicks are due to arrive. Also purchase chick starter feed and chick-sized grit ahead of time. Install new chicks in the brooder.

» **Mid-Spring.** Chicks are feathering out rapidly. They can go outside as soon as they're covered. Egg production in the older hens accelerates when they're out each day to bask in the sun and feast on bugs.

» **Late Spring.** Meat chickens should be in pens and out on pasture by now, and laying hens should be let out during the day.

» **Early Summer.** Watch for predator activity and bring chickens closer to home if some start disappearing. Order meat chicks for midsummer delivery and fall butchering.

» **Midsummer.** Butcher meat chickens started in spring. If you intend to butcher in the fall, start chicks now.

» **Late Summer.** Older hens are molting; egg laying stops. The new batch of meat chickens are out on pasture.

» **Early Fall.** Old hens come out of molt and start laying again, and new poults begin to lay. Butcher meat chickens that were started midsummer.

» **Mid-Fall.** Clean the coop and put down fresh bedding for winter. Put a heater under the water fount, and hang heat lamps if desired. Move pasture pens out of the way.

» **Late Fall.** Hang a heat lamp if desired. Check ventilation; there should be fresh air moving through, but no bad drafts.

» **Early Winter.** Collect eggs mid-morning to get them before they freeze.

Equipment Shed

» **Midwinter.** Keep any tools and equipment that are in use clean and lubricated.

» **Late Winter.** Clean and organize, and stock up on supplies. Watch for auctions and sales. As time and weather allow, check over equipment and start engines.

» **Early Spring.** Pull out tillage and planting implements so they're ready to go.

» **Mid-Spring.** Clean up and store planting equipment as you finish using it. Pull out cultivating equipment.

» **Late Spring.** Clean and store planting equipment, keep cultivators tuned up, and pull out haying equipment.

» **Early Summer.** Pull out haying equipment and make sure you have plenty of supplies and parts on hand.

» **Midsummer.** Clean up haying equipment and get it ready for the next cutting. Pull out implements for small grain harvest. Keep mower blades sharp.

» **Late Summer.** Keep haying equipment running.

» **Early Fall.** Do annual maintenance on small engines. As you finish using implements for the year, clean, lubricate, and store them.

» **Mid-Fall.** Finish cleaning, doing annual maintenance, and repairing all equipment, and store it for the season or put it somewhere handy if it will be used during winter.

» **Late Fall.** Place any equipment that will be used during winter near the shed door for easy access.

» **Early Winter.** Keep chain saws clean and blades sharpened.

Woodlot

» **Midwinter.** Winter is prime season for timber harvest and forest improvement work (pruning, thinning, debris cleanup) in regions where the ground is frozen or dry. It's also a good time for burning debris if there is snow on the ground or if the soil is wet. If you have sugar maples and plan to tap them, check over your equipment.

» **Late Winter.** Wrap up timber cutting, thinning, pruning, and burning projects for the season, in anticipation of wet and muddy conditions to come. Make maple syrup if you have sugar maples.

» **Early Spring.** Plant trees and control invasive species. Maple sugaring ends as both night and daytime temperatures rise. If there's raw soil on the trails, seed those areas down and spread a light mulch.

» **Mid-Spring.** Prime nesting season. Monitor for invasive forbs and pull them before they go to seed.

» **Late Spring.** Control invasive species.

» **Early Summer.** There's not much to do except control invasive species. You can cut timber and firewood if it's dry and there are no disease concerns. If you're mowing fence lines, mow the woods trails while you're at it — this freshens up forages on the trails for grazing wildlife and minimizes ticks on trails.

» **Midsummer.** If it's dry and you have the time, you can cut timber for firewood. Though if flies and heat are intense, why not wait for cooler weather?

» **Late Summer.** If the weather is dry, upgrade wetland crossings on woodland trails.

» **Early Fall.** As temperatures moderate and bug populations drop, cruise the woods to plan where you'll cut firewood, harvest timber, and thin and prune in the coming seasons. Mow trails one last time if needed.

» **Mid-Fall.** If there's time and the soil is soft, pull woody invasive plants. Start other woods projects and make firewood if time allows.

» **Late Fall.** Start or continue cutting firewood and thinning and pruning trees.

» **Early Winter.** Except where winters are wet rather than dry or frozen, this is the start of primary season for thinning and pruning and harvesting timber.

Wildlife Habitat

» **Midwinter.** Research what native plants and animals occupied your acres before settlement. Build birdhouses and bat and wild bee houses. Order native plant seed, food trees, and shrubs for wildlife.

» **Late Winter.** Build fish cribs (after securing a permit, if necessary) and position them on the ice so they will fall to the lake bottom when the ice melts.

» **Early Spring.** Till the soil and plant native grasses and forbs in set-aside areas as soon as the soil can be worked. Put up birdhouses built during winter. Established prairies may be burned at this season.

» **Mid-Spring.** Plant native flowers for butterflies and beneficial insects. Identify and record what native plants are already growing on your land. Plant food plots for wildlife.

» **Late Spring.** Continue observing growth of resident native species and control weeds in plantings. Avoid making hay or mowing pastures where ground-nesting birds are present.

» **Early Summer.** Prairie plantings may benefit from a high mowing if annual weeds are overtaking the slower-growing perennial natives. Mow around new trees.

» **Midsummer.** Place turtle logs and fish cribs. Put coarse rock in shallows for small fish.

» **Late Summer.** If the weather is dry, this is a good time to dig a pond. Mow forage food plots.

» **Early Fall.** Plant winter food plots.

» **Mid-Fall.** Build brush piles with debris from woods projects. Build sheltered winter feeders for ground-feeding birds.

» **Late Fall.** Build brush piles and tangles with debris from pruning, thinning, and making firewood. Leave some big logs on the ground to act as drumming logs for grouse and shelter for amphibians.

» **Early Winter.** Put out mineral and salt licks for wildlife.

Resources

Websites of Interest

ATTRA (Appropriate Technology Transfer for Rural Areas)
attra.ncat.org
Contains lots of excellent monographs on specific aspects of sustainable agriculture.

NCAT (National Center for Appropriate Technology)
ncat.org
Like ATTRA, site has many studies on specific aspects of sustainable agriculture.

HRM International (formerly Holistic Resource Management)
holisticmanagement.org
This site conveys a superb approach to creating an integrated, proactive plan for your land, including techniques for identifying and ameliorating problem areas before they become serious.

The Rodale Institute
rodaleinstitute.org
Contains decades of research and publishing on organic crop rotations, farming, and gardening techniques.

Soil

NRCS (Natural Resources Conservation Service)
nrcs.usda.gov
Home of the national soil survey.

OMRI (Organic Materials Review Institute)
omri.org
Lists products approved for organic production, including soil amendments.

Livestock

Temple Grandin's website
grandin.com
Houses detailed information on how to handle livestock humanely and calmly, and on building livestock facilities that work smoothly. It is arguably the best such information in the world.

The Livestock Conservancy
livestockconservancy.org
The best source for information on heritage breeds and what climate regions they are best adapted to.

Rotational Grazing

HRM International (formerly Holistic Resource Management)
holisticmanagement.org
This is the organization that put rotational grazing on the map by building the science behind the practice and integrating it into whole-farm systems.

Stockman Grass Farmer
stockmangrassfarmer.com
Periodical deals with all things rotational grazing; offers an excellent book store as well.

Cover Crops

SARE (Sustainable Agriculture Research and Education)
sare.org
A department of the USDA, SARE has many resources for sustainable farmers, including Managing Cover Crops Profitably.

The Xerces Society
xerces.org
Focused on the preservation of beneficial native insects, it offers regional information on plantings and habitat development.

Honey Bees

American Honey Producers Association
ahpanet.com
Provides links to state organizations and other useful information.

Invasive Plants

Invasive Plant Atlas of the United States
invasiveplantatlas.org
Offers photos of many species. If you can't identify a suspect plant, it might be better to start with a state society or your agricultural Extension service, which should be simpler and quicker.

Sustainable Forestry

Game of Logging
gameoflogging.com
Offers chainsaw safety training seminars throughout the country.

Books of Note

The Apple Grower, Michael Phillips, Chelsea Green 2005.
Pretty much covers every aspect of organic apple production.

Electric Fencing, Ann Larkin Hansen, Storey 2013.
Covers in detail the theory and practice of using electric fence.

A Landowner's Guide to Managing Your Woods, Dennis L. Waterman, Mike Severson, and Ann Larkin Hansen, Storey 2011.
Award-winning book on maintaining small-scale woods for health, wildlife, biodiversity, and timber.

Making Hay, Ann Larkin Hansen, Storey 2014.
How to cut, dry, rake, gather, and store hay.

The Organic Farming Manual, Ann Larkin Hansen, Storey 2010.
A comprehensive guide to starting and running an organic farm, with more in-depth discussions of areas covered in this book's "Topic of the Season" sections, including standard soil conservation techniques.

Week-by-Week Vegetable Gardener's Handbook, Ron Kujawski and Jennifer Kujawski, Storey 2010.
Excellent detailed time-focused guide to raising vegetables.

Acknowledgments

My heartfelt thanks to my family and to all the friends, neighbors, and relatives who have helped me in so many ways through many years of farming and reporting. Most of all I'd like to thank Sarah Guare, who babysat this book through the production process, and especially Deb Burns, who encouraged me to develop an idea into a full-blown book, and guided this journey with a sure and always friendly touch. The spreadsheet idea was brilliant, Deb.

Index

Page numbers in *italic* indicate illustrations; page numbers in **bold** indicate charts.

D

E

F

G

H

I

K

L

M